高等院校石油天然气类规划教材

资源勘查工程专业英语

（第二版·富媒体）

主　编　张树林　黄文辉　白国平
副主编　姚志刚　李少华　范广娟

石 油 工 业 出 版 社

内 容 提 要

本书涵盖地球科学概论、沉积岩与沉积相、构造地质、油气地质、油气田勘探与资源评价、油气田开发地质及能源勘探进展等内容。本书在第一版的基础上完善了经典的沉积、构造及油气地质等研究内容，加入最新油气田勘探、油气田开发及新能源进展等内容。本书共5个单元49课，每课由课文、词汇与词组组成，并在最后附有课文中常见的生词词汇及词组。同时，本书以二维码为纽带，加入了音频资源。

本书可作为高等院校资源勘查工程及地质工程本科专业英语教材，也可供相关专业的师生以及从事油气田勘探开发工作的生产和科研人员参考。

图书在版编目（CIP）数据

资源勘查工程专业英语：富媒体/张树林，黄文辉，白国平主编. —2版. —北京：石油工业出版社，2020.8

高等院校石油天然气类规划教材

ISBN 978-7-5183-4160-3

Ⅰ.①资… Ⅱ.①张…②黄…③白… Ⅲ.①地质勘探—英语—高等学校—教材 Ⅳ.①P62

中国版本图书馆CIP数据核字（2020）第142888号

出版发行：石油工业出版社

（北京市朝阳区安定门外安华里2区1号楼 100011）

网 址：www.petropub.com

编辑部：（010）64523697

图书营销中心：（010）64523633

经 销：全国新华书店

排 版：三河市燕郊三山科普发展有限公司

印 刷：北京中石油彩色印刷有限责任公司

2020年8月第2版 2020年8月第1次印刷

787毫米×1092毫米 开本：1/16 印张：12.25

字数：300千字

定价：30.00元

编写人员名单

主　编:张树林　黄文辉　白国平

副主编:姚志刚　李少华　范广娟

参编教师及单位(按姓氏笔画排序):

王喜鑫　长江大学

白国平　中国石油大学(北京)

史集建　东北石油大学

李少华　长江大学

李　斌　西南石油大学

张建国　中国地质大学(北京)

张树林　东北石油大学

范广娟　东北石油大学

姚志刚　西安石油大学

徐志明　西南石油大学

黄文辉　中国地质大学(北京)

穆娜娜　中国地质大学(北京)

魏　琳　中国地质大学(北京)

第二版前言

一本好的专业英语教材能在以下四个方面起到至关重要作用:培养“懂专业,精外语”的国际化人才;使学生具有较强的专业英语听、说、读、写、译能力;让学生专业英语翻译水平达到“信、达、雅”;提高专业英语课程教学质量。

自 2010 年 10 月至今,全国高校资源勘查工程专业及地质工程专业在本科专业英语课程教学过程中一直使用由中国石油大学(北京)、东北石油大学、西安石油大学、西南石油大学、长江大学、中国地质大学(北京)、辽宁工程技术大学 7 所高校共同编写的高等院校石油天然气类规划教材《资源勘查工程专业英语》,教材对提高师生的专业英语能力和水平起到了较大促进作用;但在使用中也发现教材存在一些缺点和不足,在近十年的教学过程中,教师、学生及现场生产和科研人员对本教材提出过许多善意的修改意见和建议;鉴于这种情况,石油工业出版社组织中国石油大学(北京)、东北石油大学、西安石油大学、西南石油大学、长江大学、中国地质大学(北京)6 所高校部分石油及地矿类专业英语骨干教师,于 2020 年 5 月中旬在网上以腾讯会议形式召开了资源勘查工程专业英语教材编审研讨会,会上各高校教师结合多年专业英语教学经验及专业英语教学发展趋势,讨论并制定了新的资源勘查工程专业英语教材编写大纲,并于 6 月底成稿。

本次教材编写共分 5 个单元 49 课,涵盖了地球科学概论、沉积岩与沉积相、构造地质与板块构造、油气地质、油气田勘探与资源评价、油气田开发地质及能源勘探进展等基本内容。第一单元为地质学基础(Geological Foundation),共计 21 课,由姚志刚、黄文辉、穆娜娜、魏琳、张建国、张树林、范广娟、史集建编写;第二单元为油气地质学(Oil and Gas Geology),共计 6 课,由徐志明、李斌编写;第三单元为油气田勘探(Exploration of Hydrocarbon Fields),共计 8 课,由白国平编写;第四单元为油气田开发(Development of Oil-Gas Fields),共计6课,由李少华、王喜鑫编写;第五单元为能源勘探进展(Progress of Energy Exploration),共计8课,由白国平、张树林、范广娟、史集建、李少华、王喜鑫、黄文辉、穆娜娜、魏琳共同编写。最后,由张树林、黄文辉、白国平进行了统稿。

本书在内容编写上,力求单词及词组覆盖面大、专业内容涉及面广,并能反映国内外最新的油气地质研究新理论与新成果;在课程体系安排上,遵从专业课教学的系统性及完整性,遵循由浅入深、循序渐进的教学规律,实现专业英语教学与专业课教学相匹配。学生在学习专业英语的同时,又学习了专业知识,达到专业与外语相互促进提高的目的。

本教材是编者多年教学成果的积累和心血的结晶,同时,在教材编写过程中参考了大量国内外相关教材、专著和文献。在此,编者向为本书编写工作提供帮助的教师、文中相关文献的作者,以及支持和帮助本教材出版发行的单位及领导表示衷心感谢!

由于水平有限,教材中难免会有不妥之处,敬请广大师生及读者批评指正,以便再版时修订。

编者

2020年6月

第一版前言

伴随着我国经济持续稳定的快速发展,国民经济对石油与天然气资源的需求日益增加。目前,中国油气资源对外依存度已超过50%。为了保证国民经济发展所需的油气资源供给,石油与天然气勘探开发必须立足国内、开拓海外。培养“懂专业,精外语”的复合型人才,让学生具备较强的专业英语听、说、读、写、译能力已成为现代高校国际化人才的重要培养目标之一。

多年来,全国高校资源勘查工程专业一直缺乏一套正式出版的专业英语统编教材。鉴于这种情况,石油工业出版社组织全国7所知名高校部分石油及地矿类专业英语骨干教师,于2009年11月在中国石油大学(北京)召开了资源勘查工程专业英语教材编审研讨会,结合各高校多年专业英语教学经验及专业英语教学发展趋势,讨论并制定了资源勘查工程专业英语教材编写大纲。

教材编写共分3个单元63课,涵盖了地球科学概论、沉积岩与沉积相、构造地质与板块构造、油气成藏机理、油气田勘探与资源评价、油气田开发地质及煤田地质等基本内容。第一单元为地质学基础(Basic Geology),共计31课,由东北石油大学、中国地质大学(北京)、西安石油大学编写;第二单元为油气地质学(Oil & Gas Geology),共计25课,由中国石油大学(北京)、西南石油大学、长江大学编写;第三单元为煤田地质学(Coal Geology),共计7课,由辽宁工程技术大学承担编写。最后,由张树林、黄文辉、白国平进行了统编统审。

本书作为本科生教材,在内容编写上,力求单词及词组覆盖面大、专业内容涉及面广,并反映国内外最新的油气地质研究新理论与新成果;在体系安排上,遵从专业课教学的系统性及完整性,遵循由浅入深、循序渐进的教学规律,实现专业英语教学与专业课教学相匹配。学生在学习专业英语的同时,又学习了专业知识,达到专业与外语相互促进提高的目的。

本教材是国内7所高校石油与地矿类专业英语教师多年教学成果的积累和心血的结晶,同时,在教材编写过程中参考了大量国内外相关教材、专著和文献。在此,编者对参与本书编写的教师、文中相关文献的作者,以及支持和帮助本教材出版发行的单位及领导表示衷心感谢。

由于水平有限,教材中难免会有不妥之处,敬请广大师生及读者批评指正,以便再版时修订。

编者

2010年6月

CONTENTS

Module 1 Geological Foundation 1

Module 1

Geological Foundation

01

Lesson 1 ▸ Geology

Earth science (also known as geoscience or Earth sciences), is an all-embracing term for the sciences related to the planet Earth. It is arguably a special case in planetary science, the Earth being the only known life-bearing planet. There are both reductionist and holistic approaches to Earth sciences. The formal discipline of Earth Sciences may include the study of the atmosphere, oceans and biosphere, as well as the solid Earth. Geology is generally categorized within the geosciences.

Geology (from the Greek, geo, "Earth" and, logos, "speech") is a natural science, which studies the geological science of the Earth's history, composition, and structure, and the associated processes. It draws upon chemistry, biology, physics, astronomy, and mathematics (notably statistics) for support of its formulations.

Geology is a very complex science. There are many branches of geology, which can be grouped under the major headings of physical and historical geology.

Physical Geology

Physical geology includes **mineralogy**, the study of the chemical composition and structure of minerals; **petrology**, the study of the composition and origin of rock; **geomorphology**, the study of the origin of landforms and theirs modification by dynamic processes; **geochemistry**, the study of the chemical composition of the Earth materials and the chemical changes that occur within the Earth and on its surface; **geophysics**, the study of the behavior of rock materials and in response to stresses and according to the principles of physics; **sedimentology**, the science of the erosion and deposition of rock particles by wind, water, or ice; **structural geology**, the study of the forces that deform the Earth's rocks and the description and mapping of deformed rock bodies; **economic geology**, the study of the exploration and recovery of natural resources, such as ores and petroleum; and engineering geology, the study of the interaction of the Earth's crust with human-made structures such as tunnel, mines, dams, bridges, and building foundations.

Historical Geology

Historical geology deals with the historical development of the Earth from the study of its rocks. They are analyzed to determine their structure, composition, and interrelationships and are examined for remains of past life. Historical geology includes **paleontology**, the systematic study of past life forms; **stratigraphy**, of layered rocks and their interrelationships; **paleogeography**, of the locations of ancient land masses and their boundaries; and geologic mapping, the superimposing of geologic information upon existing topographic maps.

Historical geologists divide all time since the information of the earliest known rocks (about 4 billion years ago) into four major divisions: the **Precambrian**, **Paleozoic**, **Mesozoic**, and **Cenozoic** eras. Each, except the Cenozoic, ended with profound changes in the disposition of the Earth's continents and mountains and was characterized by the emergence of new forms of life. Broad cyclical patterns, which run through all historical geology, include a period of mountain and continent building followed one of erosion and, in turn, by a new period of elevation.

Early Geologic Studies

Observations on Earth structure and processes were made by a number of the ancients, including Herodotus, Aristotle, Lucretius, Strabo, and Seneca. Their individual efforts in the natural history of the Earth, however, provided no sustained progress. Their major contribution is that they attributed the phenomena they observed to natural and not supernatural causes. Many of the ideas expressed by these men were not to resurface until the Renaissance. Later, Leonardo da Vinci correctly speculated on the nature of fossils as remain of ancient organisms and on the role that rivers play in the erosion of land. Agricola made a systematic study of ore deposits in the early 16th century. Robert Hooke and Nicolaus Steno both made penetrating observations on the nature of fossils and sediments.

Evolution of Modern Geology

Modern geology began in the 18th century, when field studies by the French mineralogist J. E. Guttard and others proved more fruitful than speculation. The German geologist Abraham Gottlob Werner, in spite of the many errors of his specific doctrines and the diversion of much of his energy into a fruitless controversy (in which he maintained that origin of all rocks was aqueous), performed a great service for the science by demonstrating the chronological succession of rocks.

In 1795, the Scottish geologist James Hutton laid the theoretical foundation for much of the modern science with his doctrine of **uniformitarianism**, first popularized by the British John Playfair. Largely through the work of sir Charles Lyell, this doctrine replaced the opposing one of **catastrophism**. Geology in the 19th century was influenced also by the work of Charles Darwin and enriched by the researches of the Swiss–American Louis Agassiz.

In the 20th century, geology has advanced at an ever–increasing pace. The unraveling of the mystery of atomic structure and the discovery of radioactivity allowed profound advances in many phases of geologic research. Important discoveries were made during the International Geological Year (1957—1958), when scientists from 67 nations joined force in investigating problems in all branches of geology. The systematic survey of the Earth's oceans brought radical changes in concepts of crustal evolution.

As a result of numerous flyby spacecrafts, geological studies have been extended to include remote sensing of other planets and satellites in the solar system and the Moon. Laboratory analyses of rocks samples brought back from the Moon have provided insight into the early history of near - Earth space. On - site analyses of Martian soil samples and photographic mapping of its surface have given clues about its composition and geologic history, including the possibility that Mars once had enough water to form oceans. Photographs of the many active volcanoes on Jupiter's Moon have provided clues about Earth's early volcanic activity. Geological studies also have been furthered by orbiting laboratories, such as the six launched between 1964 in the Orbiting Geophysical Observatory (OGO) series and the Polar Orbiting Geomagnetic Survey (POGS) satellite launched in 1990; remote–imaging spacecraft, such as the U. S. Landsat program (Landsat7, launched in 1999, was the most recent) and French SPOT series (SPOT4, launched in 1988, was the most recent in the program); and geological studies on space shuttle missions.

New Words and Phrases

1. mineralogy *n.* 矿物学
2. petrology *n.* 岩石学
3. geomorphology *n.* 地貌学
4. geochemistry *n.* 地球化学
5. geophysics *n.* 地球物理学
6. sedimentology *n.* 沉积学
7. structural geology 构造地质学
8. economic geology 经济地质学(矿床学)
9. paleontology *n.* 古生物学
10. stratigraphy *n.* 地层学
11. paleogeography *n.* 古地理学
12. Precambrian *n.* & *adj.* 前寒武纪(的)
13. Paleozoic *n.* & *adj.* 古生代(界)(的)
14. Mesozoic *n.* & *adj.* 中生代(界)(的)
15. Cenozoic *n.* & *adj.* 新生代(界)(的)
16. uniformitarianism *n.* 均变论
17. catastrophism *n.* 灾难论

Lesson 2 ▸ The Earth

In **geology** and **astronomy**, the Earth is the fifth largest planet of the **solar system** and the only planet definitely known to support life. **Gravitation** forces have molded the Earth, like all **celestial bodies**, into a spherical shape. However, the Earth is not an exact sphere, being slightly flattened at the poles and bulging at the **equator**, which means that the equatorial diameter of 12,756 km is larger than the polar diameter of 12,714 km. And the mass of the Earth is 6×10^{24} kg. The **altitude** of the surface features range from 11 km below to 9 km above sea level, and 70% of the surface is covered with liquid water, which makes it unique in the solar system. Additionally, Earth has an atmosphere reaching several hundred kilometers into space, consisting of 78% nitrogen, 21% oxygen, 1% other gases including carbon dioxide and water vapour. Earth is the only planet in our solar system to have large quantities of free oxygen in its atmosphere.

Knowledge of the Earth's interior has been gathered by three methods: by the analysis of earthquake waves passing through the Earth, by analogy with the composition of **meteorites**, and by consideration of the Earth's size, shape, and density. Research by these methods indicates that the Earth has a zoned interior, consisting of **concentric** shells differing from one another by size, chemical makeup, and density. The planet Earth is made up of three main shells: the very thin, **brittle** crust, the mantle, and the core; the mantle and core are each divided into two parts (Figure 2.1). Although the core and mantle are about equal in thickness, the core actually forms only 15 percent of the Earth's volume, whereas the mantle occupies 84 percent. The crust makes up the remaining 1 percent.

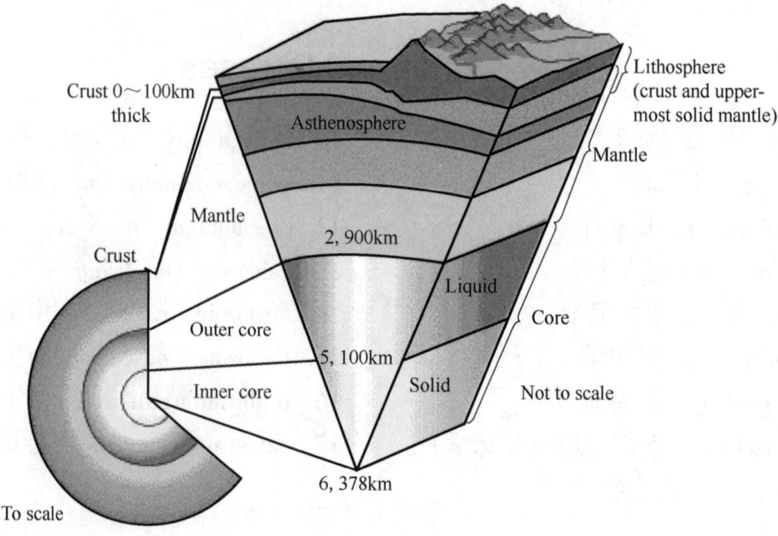

Figure 2.1 The Earth's Layers

The Earth's Crust and the Moho

Because the crust is accessible to us, its geology has been extensively studied, and therefore much more information is known about its structure and composition than about the structure and composition of the mantle and core. Within the crust, **intricate** patterns are created when rocks are redistributed and deposited in layers through the geologic processes of eruption and intrusion of **lava**, erosion, and consolidation of rock particles, and solidification and recrystallization of porous rock. By the large-scale process of plate tectonics, about twelve plates, which contain combinations of continents and ocean basins, have moved around on the Earth's surface through much of geologic time. The edges of the plates are marked by concentrations of earthquakes and volcanoes. **Collisions** of plates can produce mountains like the **Himalayas**, the tallest range in the world. The plates include the crust and part of the upper mantle, and they move over a hot, yielding upper mantle zone at very slow rates of a few centimeters per year, slower than the rate at which fingernails grow. The crust is much thinner under the oceans than under continents.

The boundary between the crust and mantle is called the Mohorovicic discontinuity (or Moho); it is named in honor of the man who discovered it, the Croatian scientist Andrija Mohorovicic. No one has ever seen this boundary, but it can be detected by a sharp increase downward in the speed of earthquake waves there. The explanation for the increase at the Moho is presumed to be a change in rock types. Drill holes to penetrate the Moho have been proposed, and a Soviet hole on the Kola Peninsula has been drilled to a depth of 12km, but drilling expense increases enormously with depth, and Moho penetration is not likely very soon.

The Earth's Mantle

Extending to a depth of about 2,900km, the mantle probably consists of very dense rock (average density about 3.9g/cm^3) rich in iron and magnesium minerals. Although temperatures increase with depth, the melting point of the rock is not reached because the melting temperature is raised by the great confining pressure. At depths between about 100km and about 200km in the mantle, a plastic zone, called the **asthenosphere**, is found to occur. Presumably the rocks in this region are very close to melting, and the zone represents a fundamental boundary between the moving **crustal plates** of the Earth's surface and the interior regions. The molten **magma** that intrudes upward into crustal rocks or issues form a volcano in the form of lava may owe its origin to radioactive heating or to the relief of pressure in the lower crust and upper mantle caused by earthquake faulting of the overlying crustal rocks. Similarly, it is thought that the heat energy released in the upper part of mantle has broken the Earth's crust into vast plates that slide around on the plastic zone, setting up stresses along the plate margins that result in the formation of folds and faults.

The Earth's Core

Though to be composed of iron and nickel, the dense core (average density: about 11.0g/cm^3) of the Earth lies below mantle. The abrupt disappearance of direct **compressional** earthquake waves, which cannot travel through liquids, at depths below about 2,900 km indicates that the outer 2,200km of the core are molten. It is thought, however, that the inner 1,260 km of the core are solid. The outer core is thought to be the source of the Earth's magnetic field: In the "**dynamo theory**" advanced by W. M. Elasser and E. Bullard, tidal energy or heat is converted to mechanical energy in the form of currents in the liquid core; this mechanical energy is then converted to electromagnetic energy, which we see as the magnetic field.

The Earth's Age

The Earth is estimated to be 4.5 billion to 5 billion years old, based on **radioactive** dating of **lunar** rocks and meteorites, which are thought to have formed at the same time. The origin of the Earth continues to be controversial. Among the theories as to its origin, the most prominent are **gravitational condensation hypotheses**, which suggest that the entire solar system was formed at one time in a single series of processes resulting in the accumulation of diffuse **interstellar** gases and dust into a solar system of discrete bodies. Old and now generally discredited theories invoked extraordinary events, such as gravitational disruption of a star passing close to the sun or the explosion of a companion star to the sun.

New Words and Phrases

1. geology *n.* 地质学
2. astronomy *n.* 天文学
3. solar system 太阳系
4. gravitation *n.* 地心吸力
5. celestial body 天体
6. equator *n.* 赤道
7. altitude *n.* (海拔)高度
8. meteorite *n.* 陨石
9. concentric *adj.* 同心的
10. brittle *adj.* 易碎的
11. intricate *adj.* 复杂的
12. collision *n.* 碰撞
13. Himalayas *n.* 喜马拉雅山脉
14. Moho *n.* 莫霍面 (Mohorovicic discontinuity 莫霍不连续面)
15. asthenosphere *n.* 软流圈
16. radioactive *adj.* 放射性的
17. magma *n.* 岩浆
18. compressional *adj.* 压缩的；纵(波)的
19. dynamo theory （地磁场成因的）发电机理论
20. lunar *adj.* 月亮的，月球的
21. gravitational condensation hypotheses （天体形成的）重力凝聚假设，也称吸积作用
22. crustal plate 地壳板块
23. interstellar *adj.* 星际的
24. lava *n.* 熔岩

Lesson 3 ▸ Geologic Time Scale

Lesson 3

Evidence from **radiometric** dating indicates that the Earth is about 4,570 million years old. Geologists have divided Earth's history into a series of time intervals. These time intervals are not equal in length like the hours in a day. Instead the time intervals are variable in length. Different **spans** of time on the time scale are usually **delimited** by major geological or **paleontological** events, such as varying rock type or fossils within the **strata** and mass extinctions. For example, the boundary between the **Cretaceous** period and the **Paleogene** period is defined by the **extinction event** that marked the demise of the **dinosaurs** and many marine species. Another example is the boundary between the **Precambrian** and the **Paleozoic** which is marked by the first appearance of animals with hard parts.

The geologic time scale was formulated during the early 1800s on the basis of information gained by relative age dating of sedimentary rocks and fossils in Europe. The largest divisions of geologic time are called eons. Eons are subdivided into eras, eras into periods, periods into **epochs**, and epochs into ages.

Eons are hundreds of millions of years in duration. The **Phanerozoic Eon** is the most recent eon and began more than 540 million years ago. **Eons** are divided into smaller time intervals known as eras. The Phanerozoic is divided into three eras: **Cenozoic**, **Mesozoic** and Paleozoic. Very significant events in Earth's history are used to determine the boundaries of the eras.

Eras are subdivided into periods. The events that bound the periods are wide-spread in their extent but are not as significant as those which bound the eras. The Paleozoic is subdivided into the **Permian**, Carboniferous, **Devonian**, Silurian, **Ordovician** and **Cambrian** periods. Each unit of the Phanerozoic intervals (541 Ma to present) and the base of the **Ediacaran** is defined by a Global Standard Section and Point (GSSP), whereas the Precambrian interval is formally subdivided by absolute age, radioactive age.

The first life in the Earth, which was probably **algae** floating in the ocean, appeared approximately 3 billion years ago. The fossil record throughout the later portion of the Precambrian is sparse and indicates that just simple life forms of plants and animals existed in the ocean. Because of the sparsity of life, there is very little organic matter preserved in Precambrian sediments and, therefore, little in the way of petroleum source rocks. Many of the Precambrian sediments have been buried deep and are metamorphosed. They offer little in the way of reservoir rocks. Therefore, no significant deposits of gas and oil are known in Precambrian rocks.

The fossil record shows that at the Paleozoic Era, a great abundance of diverse plants and animals were living in the ocean. During the Ordovician, the next period in the Paleozoic, fish evolved. Plants and animals evolved to life on the land in the next period, the Silurian.

The Mesozoic Era, starting about 250 million years ago, is known as the age of dinosaurs and **reptiles**. During the **Jurassic**, the middle period of the Mesozoic, mammals and birds evolved but were apparently dominated by the reptiles throughout the Mesozoic.

Finer subdivisions of time are possible and the periods of the Cenozoic are frequently subdivided into epochs. Subdivision of periods into epochs can be done only for the most recent portion of the geologic time scale. This is because older rocks have been buried deeply, intensely deformed and severely modified by long-term Earth processes. As a result, the history contained within these rocks can not be clearly interpreted.

The Cenozoic Era is divided into three periods, the **Paleogene**, **Neogene** and **Quaternary**, and they are in turn divided into epochs. The Paleogene consists of the Paleocene, **Eocene**, and Oligocene epochs, the Neogene consists of the Miocene, Pliocene, and the Quaternary consists of Pleistocene and Holocene epochs, the last of which is ongoing. Historically, the Cenozoic Era had been divided into periods named the **Tertiary** (Paleocene through Pliocene) and **Quaternary** (Pleistocene and Holocene).

New Words and Phrases

1. radiometric *adj.* 放射测量的
2. span *n.* 跨度，范围
3. delimit *vt.* 定界线，划界；为……划界
4. paleontological *adj.* 古生物(学)的
5. strata *n.* 地层
6. Cretaceous *n.* & *adj.* 白垩纪(的)
7. Paleogene *n.* & *adj.* 古近纪(的)
8. extinction event （生物)灭绝事件
9. dinosaur *n.* 恐龙
10. Precambrian *n.* & *adj.* 前寒武(的)
11. Paleozoic *n.* & *adj.* 古生代(的)
12. eon *n.* 永世；[地]宙
13. era *n.* 时代；纪元 [地]代
14. epoch *n.* 新纪元；[地]世
15. Phanerozoic *adj.* 显生宙的
16. Cenozoic *n.* & *adj.* 新生代(的)
17. Mesozoic *n.* & *adj.* 中生代(的)
18. Permian *n.* & *adj*. 二叠纪(的)
19. Devonian *n.* & *adj*. 泥盆纪(的)
20. Ordovician *n.* & *adj*. 奥陶纪(的)
21. Cambrian *n.* & *adj*. 寒武纪(的)
22. Ediacaran *n.* 埃迪卡拉纪
23. algae *n.* 藻类，海藻
24. reptile *n.* & *adj*. 爬虫动物(的)
25. Jurassic *n.* & *adj*. 侏罗纪(的)
26. Neogene *n.* & *adj*. 新近纪(的)
27. Paleocene *n.* & *adj*. 古新世(的)
28. Eocene *n.* & *adj*. 始新世(的)
29. Tertiary *n.* & *adj*. 古近—新近纪(的)
30. Quaternary *n.* & *adj*. 第四纪(的)

Lesson 4 ▸ Common Minerals

Lesson 4

Mineral is inorganic substance occurring in nature, having a characteristic and **homogeneous** chemical composition, definite physical properties, and usually, a definite **crystalline** form. A few of the minerals (e. g. , carbon, **arsenic**, **bismuth**, **antimony**, gold, silver, copper, **lead**, **mercury**, **platinum**, and iron) are elements, but the vast majority are chemical compounds. A generalized formula can usually be assigned to each mineral that is a chemical compound, although sometimes one element in a mineral may be replaced by another without changing the species of the mineral (**isomorphism**). Minerals combine with each other to make up rocks, which, as distinguished from minerals, are of **heterogeneous** composition. Minerals may occur in the massive state when conditions for the formation of crystals are unfavorable. Among the important physical properties of minerals are specific gravity, hardness, cleavage, fracture, luster, color, **transparency**, streak, striations, tenacity, **fusibility**, **heat conductivity**, taste, odor, feel, magnetism, and electrical properties. Minerals originate by precipitation from solution, by the cooling and hardening of magmas, by the condensation of gases or gaseous action on country rock, and by **metamorphism**. Minerals in rocks are frequently replaced by other minerals through the action of water or gases (metasomatism). Minerals, especially the metals, are of great economic importance to a highly industrialized civilization, entering into the composition of many manufactured articles. Many minerals which would otherwise be of no economic significance are highly valued as gems. Mineralogy, a branch of geology, is the science of minerals.

Ore minerals are economically valuable minerals that through various methods of mining and mineral extraction can be separated from waste and processed to yield valuable **commodities**. This definition is used, in the strictest sense, by economic geologists and mining engineers to apply only to those minerals that contain the valuable commodity of interest. For example, they do not consider that worthless iron **sulphides** associated with valuable copper sulphides in a copper deposit are ore minerals. However, the term "ore minerals" is also encountered in **petrography**, where it refers to any mineral that is opaque to transmitted light when examined in thin section. Ore minerals are found in several of the main chemical classes into which minerals can be classified, including native elements, sulphides, oxides, **silicates**, **arsenides**, **phosphates**, and carbonates.

The simplest ore minerals are the native metals in which the mineral is composed of a single element. The best-known example in this group is native gold, which is the most common ore mineral of this metal. Other metals for which the native form is an important ore mineral include silver, copper, and bismuth. Two or more metals can also combine to form a group of ore minerals called **alloys**. **Electrum**, an alloy of gold and silver, is a common mineral in gold-silver deposits, and many of the metals of the platinum group combine with each other, as well as with other metals, to form complex alloys.

The sulphide minerals comprise an important group of ore minerals in which one or more metals combine with sulphur. Simple sulphide minerals are formed when a single metal combines with sulphur. For example, **sphalerite** and **galena** (the simple sulphides of zinc and lead), are the world's primary source for these metals. Other rarer simple sulphides are important sources of antimony, arsenic, bismuth, copper, mercury, molybdenum, nickel, and silver. Sulphide minerals in which two metals combine with sulphur are also common ore minerals. **Chalcopyrite**, a combination of copper, iron, and sulphur, is an important ore mineral for copper and is found in many of its deposits. Other more complex sulphides, formed by the combination of two or more metallic or metalloid elements with sulphur, are called sulphosalts. These minerals are less common than the simpler sulphides and are important as ore minerals in only a few deposits.

Many oxide minerals also constitute economically important ore minerals. Simple oxides, in which a single metal combines with oxygen, are the major ore minerals of iron (magnetite and haematite), tin (cassiterite), manganese (pyrolusite), titanium (rutile), and uranium (uraninite). Hydrated oxides, in which water is incorporated into the chemical structure of the mineral, also provide important sources for several metals, including iron (goethite) and manganese (manganite). **Bauxite**, the major ore of **aluminum**, is composed of a mixture of hydrated aluminum oxides (gibbsite, diaspore, boehmite).

Although many metals combine with silicon and oxygen to form silicate minerals, these cannot in general be refined economically because of the large amounts of energy required to extract the metals. There are, however, some metals for which silicate minerals are an important source. For example, the major ore minerals for lithium and beryllium (spodumene and beryl) are silicate minerals.

The **halides** are a group of minerals whose principle anions are halogens. The halogens that are found commonly in nature include fluorine, chlorine, iodine and bromine. Halides tend to have rather simply ordered structures and therefore a high degree of symmetry. The most famous halide mineral, halite or rock salt has the highest symmetry. The colorful mineral fluorite has cubic crystals, being very popular mineral specimens. The typical halide mineral is soft, can be transparent, is generally not very dense, has a good cleavage, and often has bright colors.

New Words and Phrases

1. homogeneous *adj.* 同类的，均质的
2. crystalline *adj.* 晶体的，结晶的
3. arsenic *n.* ［化］砷（符号 As）；砒霜
4. bismuth *n.* ［化］铋（符号 Bi）
5. antimony *n.* ［化］锑（符号 Sb）
6. lead *n.* ［化］铅（符号 Pb）

7. mercury *n.* 水银，汞(符号 Hg)
8. platinum *n.* 白金，铂(符号 Pt)
9. isomorphism *n.* 类质同象
10. heterogeneous *adj.* 异种的，非均质的
11. transparency *n.* 透明度
12. fusibility *n.* 熔点
13. conductivity *n.* 传导率
    (Heat conductivity 热传导率)
14. metasomatism *n.* [地]交代(作用)，交代变质(作用)
15. commodity *n.* 日用品，商品
16. sulphide *n.* 硫化物
17. petrography *n.* 岩相学
18. silicate *n.* 硅酸盐
19. arsenide *n.* 砷化物
20. phosphate *n.* 磷酸盐
21. alloy *n.* 合金
22. electrum *n.* 银金矿,金银合金
23. sphalerit *n.* 闪锌矿
24. galena *n.* 方铅矿
25. chalcopyrite *n.* 黄铜矿
26. bauxite *n.* 铝矾土，铝土矿
27. aluminum *n.* 铝(符号 Al)
28. halides *n.* 卤化物

Lesson 5

Lesson 5 ▸ Rocks

Rock is **aggregation** of solid matter composed of one or more of the minerals forming the Earth's crust. The scientific study of rocks is called **petrology**. Rocks are commonly divided, according to their origin, into three major classes—**igneous**, **sedimentary**, and **metamorphic**.

Igneous Rocks

Igneous rock originates from the cooling and **solidification** of molten matter from the Earth's interior. If the rock is formed on the Earth's surface (i. e., from the solidification of lava), it is called **extrusive rock**; **igneous rock** that has cooled and solidified slowly beneath the Earth's surface is intrusive rock. Among the forms commonly taken by intrusive rocks are batholiths, which are enormous, irregular masses cutting or displacing older rocks; stocks, irregular and smaller than batholiths; necks, or plugs, columnar in form and probably the result of the hardening of magma in the necks of extinct volcanoes; dikes, more or less vertical, filling fissures in previously existing rock; sills, more or less horizontal, forced between layers of previously existing rock; and laccoliths, modified domelike sills that arch under the overlying rock.

Igneous rocks are commonly divided into classes by texture. Some rocks are markedly **granular** (e. g., granite, syenite, diorite, gabbro, peridotite, and pyroxenite), while others (e. g., basalt, trachite, dacite, and andesite) are composed of grains visible only under a microscope. Both fine-grained and coarse-grained igneous rocks frequently contain grains called phenocrysts that are larger than the surrounding grains; such rocks are said to be porphyritic in texture. Rocks with grains of uniform size are called equigranular.

Igneous rocks are commonly light in color if their constituent minerals are predominantly alkali **feldspars** and dark in color if the feldspars are calcic or if magnesia and iron minerals are abundant. The glassy igneous rocks include obsidian, pitchstone, and pumice, which contain few or no phenocrysts, and vitrophyre, or glass porphyry, which does contain phenocrysts. Rocks such as tuff and volcanic breccia, which are formed from fragmental volcanic material, are sometimes grouped as pyroclastic rocks.

Sedimentary Rocks

Sedimentary rocks originate from the consolidation of sediments derived in part from living organisms but chiefly from older rocks of all classes (ultimately the mineral elements are derived from igneous rocks alone). The sediments of inorganic origin are chiefly removed from older rocks by erosion and transported to the place of deposition; chemical precipitation from solution is a secondary cause of deposition of inorganic matter. Sedimentary rocks are commonly distinguished, according to their place of deposition, by a great variety of terms, such as continental, marine (i. e., oceanic), littoral (i. e., coastal), estuarine (i. e., in an estuary), lacustrine (i. e., lakes), and fluviatile, or fluvial (i. e., in a stream).

The characteristic feature of sedimentary rocks is their **stratification**; they are frequently called stratified rocks. Sedimentary rocks made up of angular particles derived from other rocks are said to have a **clastic** texture, in contrast to pyroclastic sediments, which are particles of volcanic origin. Among the important varieties of sedimentary rock, distinguished both by texture and by chemical composition, are conglomerate, sandstone, tillite, sedimentary breccia, shale, marl, chalk, limestone, coal, lignite, gypsum, and rock salt. Characteristic occurrences in sedimentary rocks are fossils, footprints, raindrop impressions, concretions, oolites, ripple marks, rill marks, and **crossbedding**. Some of these features are useful in determining the antiquity of sedimentary formations and in interpreting geologic history.

Metamorphic Rocks

Metamorphic rocks originate from the alteration of the **texture** and mineral constituents of igneous, sedimentary, and older metamorphic rocks under extreme heat and pressure deep within the Earth. Some (e. g., marble and quartzite) are massive in structure; others, and particularly those which have been subject to the more extreme forms of metamorphism, are characterized by foliation (i. e., the arrangement of their minerals in roughly parallel planes, giving them a banded appearance). A distinguishing characteristic of many metamorphic rocks is their **slaty cleavage**. Among the common metamorphic rocks are schist (e. g., mica schist and hornblende schist), gneiss, quartzite, slate, and marble.

Metamorphic rocks can generally be divided into four types, according to the heat and pressure they had been undergone: (1) regional metamorphic rocks, formed in response to changes leading to high temperature and high pressure (**shearing stress** and **hydrostatic pressure** accompanying orogenic events; (2) contact metamorphic rocks, formed in response to changes leading to high temperature (with low pressure) around an igneous intrusion; (3) cataclastic or dynamic metamorphic rocks, formed in response to an increase in directed pressure (shearing stress) particularly in fault and thrust zones; (4) burial metamorphism, formed in response to changes leading to high pressure (with low temperature).

New Words and Phrases

1. aggregation *n.* 集合，集合体
2. petrology *n.* 岩石学
3. igneous *adj.* 火的；[地]火成的
4. sedimentary *adj.* 沉积的
5. metamorphic *adj.* 变质的
6. solidification *n.* 凝固，固化
7. extrusive rock 喷出岩
8. intrusive rock 侵入岩

9. granular *adj.* 粒状的
10. feldspar *n.* 长石
11. stratification *n.* 层理
12. clastic *adj.* 碎屑(状)的
13. crossbedding *n.* 交错层理
14. texture *n.* 结构
15. slaty cleavage 板状劈理;板岩劈理
16. shearing stress 剪应力
17. hydrostatic pressure 静水压力

Lesson 6 ▸ Stratigraphy and Paleontology

Stratigraphy

Stratigraphy, a branch of geology, studies rock layers and layering (stratification). It is primarily used in the study of sedimentary and layered volcanic rocks. The **theoretical** basis for the subject was established by Nicholas Steno who reintroduced the **law of superposition** and introduced the principle of original horizontality and principle of lateral continuity in a 1669 work on the fossilization of organic remains in layers of sediment.

The first practical large scale application of stratigraphy was by William Smith in the 1790s and early 1800s. Smith, known as the Father of English Geology, created the first geologic map of England, and first recognized the significance of strata or rock layering, and the importance of fossil markers for correlating strata. Another influential application of stratigraphy in the early 1800s was a study by Georges Cuvier and Alexandre Brongniart of the geology of the region around Paris. Stratigraphy includes two related subfields: **lithostratigraphy** or lithologic stratigraphy and biologic stratigraphy or **biostratigraphy**.

Lithostratigraphy, or lithologic stratigraphy, is the most obvious. It deals with the physical lithologic, or rock type, changing both vertically in layering or bedding of varying rock type and laterally reflecting changing **environments of deposition**, known as facies change. Key elements of stratigraphy involve understanding how certain **geometric** relationships between rock layers arise and what these geometries mean in terms of depositional environment. One of stratigraphy's basic concepts is codified in the Law of Superposition, which simply states that, in an undeformed stratigraphic sequence, the oldest strata occur at the base of the sequence.

Chemostratigraphy is based on the changes in the relative proportions of trace elements and isotopes within and between lithologic units. Carbon and oxygen isotope ratios vary with time and are used to map subtle changes in the paleoenvironment. This has led to the specialized field of isotopic stratigraphy.

Cyclostratigraphy documents the often cyclic changes in the relative proportions of minerals, particularly carbonates, and fossil diversity with time, related to changes in palaeoclimates.

Biologic stratigraphy or biostratigraphy is based on fossil evidence in the rock layers. Strata from widespread locations containing the same fossil **fauna** and **flora** are correlatable in time. Biologic stratigraphy was based on William Smith's principle of **faunal succession**, which predated, and was one of the first and most powerful lines of evidence for, biological evolution. It provides strong evidence for formation (**speciation**) of and the extinction of species. The geologic time scale was developed during the 1800s based on the evidence of biologic stratigraphy and faunal succession. This timescale remained a relative scale until the development of radiometric dating, which gave it and the stratigraphy based on an absolute time framework, leading to the development of chronostratigraphy.

One important development is the **Vail curve**, which attempts to define a global historical sea-level curve according to inferences from world-wide stratigraphic patterns. Stratigraphy is also commonly used to delineate the nature and extent of hydrocarbon-bearing reservoir rocks, seals and traps in petroleum geology.

Paleontology

Paleontology is the study of **prehistoric** life, including organisms' evolution and interactions with each other and their environments (their **paleoecology**). As a "historical science", it tries to explain causes rather than conduct experiments to observe effects. Paleontological observations have been documented as far back as the fifth century BC. The science became established in the 18th century as a result of Georges Cuvier's work on **comparative anatomy**, and developed rapidly in the 19th century. Fossils found in China since the 1990s have provided new information about the earliest evolution of animals, early fish, dinosaurs and the evolution of birds and mammals. Paleontology lies on the border between biology and geology, and shares with **archaeology** a border that is difficult to define. It now uses techniques drawn from a wide range of sciences, including biochemistry, mathematics and engineering.

Body fossils and trace fossils are the **principal** types of evidence about ancient life, and geochemical evidence has helped to **decipher** the evolution of life before there were organisms large enough to leave fossils. Estimating the dates of these remains is essential but difficult: sometimes adjacent rock layers allow radiometric dating, which provides absolute dates that are accurate to within 0.5%, but more often paleontologists have to rely on relative dating by solving the "**jigsaw puzzles**" of biostratigraphy. Classifying ancient organisms is also difficult, as many do not fit well into the **Linnean taxonomy** that is commonly used for classifying living organisms, and paleontologists more often use **cladistics** to draw up evolutionary "family trees". The final quarter of the 20th century saw the development of molecular **phylogenetics**, which investigates how closely organisms are related by measuring how similar the DNA is in their **genomes**. Molecular phylogenetics has also been used to estimate the dates when species diverged, but there is controversy about the reliability of the molecular clock on which such estimates depend.

Use of all these techniques has enabled paleontologists to discover much of the evolutionary history of life, almost all the way back to when Earth became capable of supporting life, about 3,800 million years ago. For about half of that time the only life was single-celled micro-organisms, mostly in microbial mats that formed ecosystems only a few millimeters thick. Earth's atmosphere originally contained virtually no oxygen, and its oxygenation began about

2,400 million years ago. This may have caused an accelerating increase in the diversity and complexity of life, and early multicellular plants and **fungi** have been found in rocks dated from 1,700 to 1,200 million years ago. The earliest multicellular animal fossils are much later, from about 580 million years ago, but animals diversified very rapidly and there is a lively debate about whether most of this happened in a relatively short *Cambrian explosion* or started earlier but has been hidden by lack of fossils. All of these organisms lived in water, but plants and invertebrates started colonizing land from about 490 million years ago and vertebrates followed them about 370 million years ago. The first dinosaurs appeared about 230 million years ago and birds evolved from one dinosaur group about 150 million years ago. During the time of the dinosaurs, mammals' **ancestors** survived only as small, mainly nocturnal insectivores, but after the non-avian dinosaurs became extinct in the Cretaceous-Pliocene extinction event 65 million years ago mammals diversified rapidly. Flowering plants appeared and rapidly diversified between 130 million years ago and 90 million years ago, possibly helped by **coevolution** with **pollinating** insects. The upright-walking common **ancestor** of humans appeared around 6 million to 7 million years ago, and anatomically modern humans appeared under 200,000 years ago. The course of evolution has been changed several times by mass extinctions that wiped out previously dominant groups and allowed other to rise from obscurity to become major components of ecosystems.

New Words and Phrases

1. stratigraphy *n.* 地层学
2. theoretical *adj.* 理论的
3. law of superposition 地层层序律
4. lithostratigraphy *n.* 岩石地层学
5. biostratigraphy *n.* 生物地层学
6. environments of deposition 沉积环境
7. geometric *adj.* 几何(学)的
8. fauna *n.* 动物群
9. flora *n.* 植物群
10. faunal succession 动物区系演替
11. speciation *n.* 物种形成
12. chronostratigraphy *n.* 地质年代学
13. Vail curve 凡尔(海平面)曲线
14. paleontology *n.* 古生物学
15. prehistoric *adj.* 史前的,陈旧的
16. paleoecology *n.* 古生态学
17. comparative anatomy 比较解剖学
18. archaeology *n.* 考古学
19. principal *adj.* 首要的,主要的
20. decipher *vt.* 译解,解释
21. jigsaw puzzles 智力拼图游戏
22. Linnean taxonomy 林氏分类法
23. cladistics *n.* 遗传分类学
24. phylogenetics *n.* 系统发生学
25. genome *n.* 基因组,染色体组
26. fungi *n.* 真菌类
27. colonize *n.* 拓殖,殖民
28. ancestor *n.* 祖先,祖宗
29. coevolution *n.* 共进化
30. pollinate *vt.* 对……授粉

Lesson 7 ▸ The Earth's Resource

In classifying natural resources, it has been traditional to distinguish between those that are **renewable** and those that are non-renewable. The former were once considered to be the living resources, e. g. , forests, wildlife, and so on. Because of their ability to regenerate through reproduction. The latter were considered to be nonliving mineral or fuel resources, which, once used, did not replace themselves.

In practice, this separation is not entirely satisfactory. At the extreme, all resources on this planet (except sunlight) are finite, and therefore in theory can be exhausted. In practice, however, most **non-renewable** resources (metals, minerals), with the exception of fossil fuels, are not destroyed by their usage, and with efficient recycling, these materials can be "renewed". They are only considered "non-renewable" in the sense that natural formation of new economic deposits of these materials occurs on timescales that greatly exceed human experience (e. g. , tens of thousands to millions of years). For most non-renewable materials, supply has kept up with demand over the last century, new resources are continually being discovered, and new technologies increase the efficiency of mineral extraction and processing; hence, in the short to medium term, future generations are not being disadvantaged by our current usage. In the longer term, however, easily mined (near surface, high grade) resources will become harder to find. This will cause price increases, which in turn will mean that lower grade and harder-to-reach deposits will become economic to mine. Substitution by other cheaper materials may also occur, and recycling will increase as the value of scrap rises.

There are, nevertheless, certain aspects of conservation that apply specifically to nonliving resources:

(1) **Beneficiation** is the upgrading of a resource that was once too uneconomical to develop. It usually depends upon technological improvements, such as those that make possible the concentration of a dispersed fuel or mineral so that it can be more easily handled, transported, or processed.

(2) Maximization is the aggregate of those measures that avoid waste and increase the production of a resource.

(3) Substitution involves the use of common resources in place of rare ones, as, for example, the use of aluminum in place of less abundant copper for a variety of products.

(4) **Allocation** is the determination of the most appropriate use for a resource and the assignment of the resource to that purpose. In market economies, allocation is usually controlled by the pricing mechanism: if the demand for a particular purpose is high, then the price of a resource used for that purpose will also be high; this high price will in turn make it more likely that the resource will be used mostly for that purpose. In government-controlled economies, a

resource may be reserved only for what are considered to be its most important uses.

(5) **Recycling**, one of the most promising methods for conservation of mineral resources, involves the concentration of used or waste materials, their reprocessing (if this is required), and their subsequent reutilization in place of new materials. If carried out in an organized and consistent manner, recycling can greatly reduce the drain on supplies of minerals. It is also appropriate for products derived from living resources, such as the reuse of wood and paper as well as the reclamation of organic fertilizers from **sewage**.

In contrast, it is many of the " renewable" resources that face the greatest risk of exhaustion, and in the case of living resources (plants, animals), potential extinction. Renewable resource depletion, such as deforestation, over-fishing, soil degradation, and river and groundwater drawdown, can cause long-term damage to eco-systems far in excess of that caused by extraction and use of non-renewable resources, with the exception of changes to the Earth's atmosphere from the burning of fossil fuels.

Because all natural resources form a continuum, from those that are most renewable in the short term to those that are least renewable, they do not readily lend themselves to a single system of classification. It is useful, therefore, to examine the various types of natural resources in relation to their cycling time; i. e. , the length of time required to replace a given quantity of a resource that has been utilized with an equivalent quantity in a similarly useful form. Form this point of view, renewable resources can be considered as those with short cycling times and nonrenewable resources as those with very long cycling times. Any resource can be nonrenewable, however, if the demand and rate of utilization exceed its cycling capacity.

Two kinds of natural resources, pasture grass and coal, can be used to illustrate the concept of cycling time. When grass is grazed by livestock or mowed, a crop of it is removed. If **provision** is made to protect the fertility and structure of the soil and to leave enough seed or adequate roots and vegetative parts to produce new growth, then a grass crop can be removed from a pasture each year for an indefinite period of time. Removal of one year's crop does not diminish the supply available for the next year if the land is cared for properly. The cycling time for this resource may be one year in areas in which climate limits growth, or it may be less than a year if growth can be continuous.

By contrast, the coal resources of the Earth were built up over millions of years. Most were laid down during the **Carboniferous** Period of geologic time (from 345Ma to 280Ma ago), when climates were warm. Extensive swamp forests covered large areas of the Earth, and conditions were favorable for plant debris to accumulate in extensive deposits without decomposing and breaking down organically. Subsequently, heart and pressure generated by the deposition of other materials on top of the organic debris and by movements of the Earth's crust transformed the plant remains into coal. Organic debris is still being produced in swaps and marshes, and over millions of years this could become transformed into coal. The time scale is so great, however, that, for human purposes, coal can be considered as a nonrenewable resource. Thus, only the supplies presently in available in the Earth′s crust can be counted on for future use.

New Words and Phrases

1. renewable *adj.* 可再生的
2. non-renewable *adj.* 不可再生的
3. beneficiation *n.* 选矿
4. allocation *n.* 分配，配置
5. recycling *n.* 重复使用，回收
6. reclamation *n.* 回收
7. sewage *n.* 污水，垃圾
8. provision *n.* 供应
9. Carboniferous *n.* & *adj.* 石炭纪(的)

Lesson 8 ▸ Sedimentary Rocks

Lesson 8

Geologists disagree somewhat about how the various kinds of sedimentary rocks should be classified; however, such rocks can conveniently be placed into three fundamental groups on the basis of composition and origin: siliciclastic, chemical/biochemical, and carbonaceous. **Siliciclastic sedimentary rocks** are composed dominantly of silicate minerals, such as quartz and feldspar, and rock fragments (clasts). These materials originate mainly by the chemical and physical breakdown (weathering) of igneous, metamorphic, or (older) sedimentary rock. **Conglomerates**, **sandstones**, and **shales** belong to this group. Silicate detritus, including silicate minerals, rock fragments, and glass shards, can also be generated by explosive volcanism. Siliciclastic sedimentary rocks that formed mainly from the products of explosive volcanism are called volcaniclastic rocks. Chemical/biochemical sedimentary rocks are composed of minerals precipitated mainly from ocean or lake water by inorganic (chemical) and/or organic (biogenic) processes. They include limestone, chert, **evaporates** such as gypsum, phosphorites, and iron-rich sedimentary rocks. Evaporites are probably precipitated entirely by inorganic processes resulting from evaporation of lake or seawater. Biogenic processes, as well as inorganic processes, play an important role in the formation of many limestones and likely play some role in the origin of chert, phosphorites, and iron-rich sedimentary rocks. Carbonaceous sedimentary rocks contain a substantial amount (>15%) of highly altered remains of the soft tissue of plants and animals, referred to as organic matter. The principal carbonaceous rocks are **coal** and oil shale. Carbonaceous sedimentary rocks make up only a small fraction of the total sedimentary record; however, these rocks (especially coals) have great economic importance as **fossil fuels**.

Sedimentary rock is a type of rock that is formed by sedimentation of material at the Earth's surface and within bodies of water. Sedimentation is the collective name for processes that cause mineral and/or organic particles (detritus) to settle and accumulate or minerals to precipitate from a solution. Particles that form a sedimentary rock by accumulating are called sediment. Before being deposited, sediment was formed by weathering and **erosion** in a source area, and then transported to the place of deposition by water, wind, mass movement or glaciers. The type of sediment that is transported to a place depends on the geology of the hinterland (the source area of the sediment). However, some sedimentary rocks, like evaporites, are composed of material that formed at the place of deposition. The nature of a sedimentary rock therefore not only depends on sediment supply, but also on the sedimentary depositional environment in which it formed.

The sedimentary rock cover of the continents of the Earth's crust is extensive, but the total contribution of sedimentary rocks is estimated to be only 5% of the total volume of the crust.

Sedimentary rocks are only a thin veneer over a crust consisting mainly of igneous and metamorphic rocks.

Sedimentary rocks are deposited in strata that form a structure called bedding. The study of sedimentary rocks and rock strata provides information about the subsurface that is useful for civil engineering, for example in the construction of roads, houses, tunnels canals or other constructions. Sedimentary rocks are also important sources of natural resources like coal, fossil fuels, drinking water or ores.

The study of the sequence of sedimentary rock strata is the main source for scientific knowledge about the Earth′s history, including **palaeogeography**, **paleoclimatology** and the history of life.

The scientific discipline that studies the properties and origin of sedimentary rocks is called sedimentology. Sedimentology is both part of geology and physical geography and overlaps partly with other disciplines in the Earth sciences, such as pedology, geomorphology, **geochemistry** or **structural geology**.

Sedimentary rocks are classified into three groups. These groups are clastic, biogenic, chemical sedimentary rocks (Table 8. 1).

Table 8. 1 Principal Sedimentary Rock Categories

Categories	Subcategories
Clastic (detrital or terrigenous) sedimentary rocks	Conglomerate and **breccia**
	Sandstone
	Mudrock (claystone, **siltstone**, mudstone, and shale)
Biogenic (biochemical or organic) sedimentary rocks	Carbonate rock (limestone and dolostone)
	Chert (silicate rock)
	Phosphate rock
Chemical sedimentary rocks	Evaporite
	Precambrian iron formation
	Phanerozoic ironstone

Clastic sedimentary rocks are composed of discrete fragments or clasts of materials derived from other minerals. They are composed largely of quartz with other common minerals including **feldspar**, amphiboles, **clay** minerals, and sometimes more **exotic** igneous and metamorphic minerals.

Biogenic sedimentary rocks contain materials generated by living organisms, and include carbonate minerals created by organisms, such as corals, mollusks, and **foraminifera**, which cover the ocean floor with layers of calcium **carbonate** which can later form **limestone**. Other examples include **stromatolites**, the flint nodules found in chalk (which is itself a biochemical sedimentary rock, a form of limestone), and coal and **oil shale** (derived from the remains of tropical plants and subjected to heat).

Chemical sedimentary rocks form when minerals in solution become **supersaturated** and precipitate. In marine environments, this is a method for the formation of limestone. Another common environment in which chemical sedimentary rocks form is a body of water that is evaporating. Evaporation decreases the amount of water without decreasing the amount of dissolved material. Therefore, the dissolved material can become oversaturated and precipitate. Sedimentary rocks from this process can include the evaporite minerals **halite** (rock salt), sylvite, barite and **gypsum**.

New Words and Phrases

1. siliciclastic sedimentary rock 硅质碎屑沉积岩
2. conglomerate *n.* 砾岩
3. coal *n.* 煤
4. oil shale 油页岩
5. fossil fuel 化石燃料
6. erosion *n.* 侵蚀
7. evaporate *n.* 蒸发岩(盐)
8. palaeogeography *n.* 古地理学
9. paleoclimatology *n.* 古气候学
10. geochemistry *n.* 地球化学
11. structural geology 构造地质学
12. clastic *adj.* 碎屑(状)的
13. feldspar *n.* 长石
14. clay *n.* 黏土
15. exotic *adj.* 外来的
16. limestone *n.* 石灰岩
17. sandstone *n.* 砂岩
18. siltstone *n.* 粉砂岩
19. breccia *n.* 角砾岩
20. shale *n.* 页岩
21. carbonate *n.* 碳酸盐岩
22. foraminifera *n.* 有孔虫类
23. stromatolite *n.* 叠层(石)
24. supersaturate *vt.* 过饱和
25. halite *n.* 岩盐
26. gypsum *n.* 石膏

Lesson 9 ▶ Weathering, Erosion, Transportation and Deposition

Weathering and Erosion

Sedimentary rocks form through a complex set of processes that begin with **weathering**, the physical disintegration and chemical decomposition of older rock to produce solid particulate residues (resistant minerals and rock fragments) and dissolved chemical substances. Weathering involves chemical, physical, and biological processes, although chemical processes are by far the most important. A brief summary of weathering processes is presented here (Figure 9.1) to illustrate how weathering acts to decompose and disintegrate exposed rocks, producing particulate residues and dissolved constituents. These weathering products are the source materials of soils and sedimentary rocks; thus, weathering constitutes the first step in the chain of processes that produce sedimentary rocks.

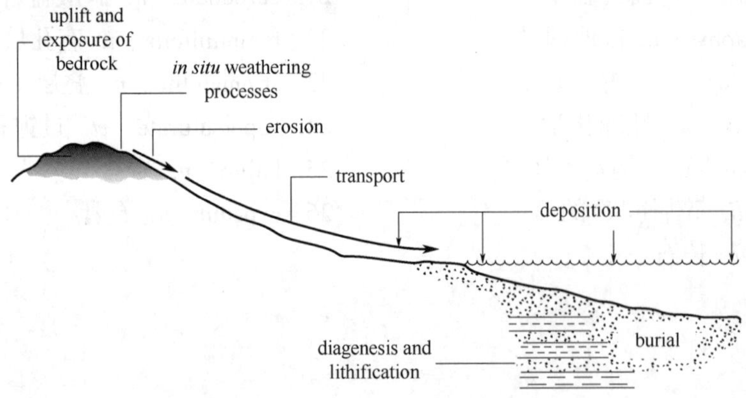

Figure 9.1 The pathway of processes involved in the formation of a succession of clastic sedimentary rocks, part of the rock cycle.

It is important to understand how weathering attacks exposed source rocks and what remains after weathering to form soils and be transported as sediment and dissolved constituents to depositional basins. The ultimate composition of soil and terrigenous sedimentary rock bears a relationship to the composition of their source rock; however, study of residual soil profiles shows that both the mineral composition and the bulk chemical composition of soils may differ greatly from those of the bedrock on which they form. Some minerals in the source rock are destroyed completely during weathering, whereas more chemically resistant or stable minerals are loosened from the fabric of the decomposing and disintegrating rock and accumulate as residues. During this process, new minerals such as iron oxides and clay minerals may form in situ in the soils from chemical elements released during breakdown of the source rocks. Thus, soils are composed of survival assemblages of minerals and rock fragments derived from the parent rocks plus any new minerals formed at the weathering site. Soil composition is governed not only by the parent-rock composition but also by the nature, intensity; and duration of weathering and soil-forming processes. It follows from this premise that the composition of terrigenous sedimentary rocks such as sandstones, which are derived from soils and other weathered materials, is also controlled by parent-rock composition and weathering processes.

Most ancient soils were probably eroded and their constituents transported to furnish the materials of sedimentary rocks; however, some survived to become part of the geologic record. We call these ancient soils—paleosols. Weathering and soil-forming processes are significantly influenced by climatic conditions. Geologists are greatly interested in the study of past climates, called paleoclimatology, because of this relationship and because paleoclimates also influenced past sea levels and sedimentation processes as well as the life forms on Earth at various times.

Transportation and Deposition

The transportation and deposition of sediments are governed by the laws of physics. Transportation is the process by which weathered material is removed from the weathering site and carried to its area of deposition. Sedimentation is, literally, the settling out of solid matter in a liquid. However, sedimentary processes are generally understood as those which both transport and deposit sediment. They include the work of water, wind, ice and gravity.

One of the most important processes for transporting and depositing sediment is **traction** currents. They move sediment along by rolling and saltation as **bed load** in a traction carpet. Continuous reworking **winnows** out the silt and clay particles which are carried off in **suspension**. Finer, lighter sand grains are transported faster than larger heavier ones. The sediments of a unidirectional traction current thus tend to show a downstream decrease in grain size, termed "size grading". Traction currents may be **unidirectional**, as in river channels. In **estuaries**, and in the open sea, sediment may be subjected to the to-and-fro action of tidal traction currents, or to even more complex systems.

Turbidity currents are widely believed to be a major process for the transportation and deposition of a significant percentage of the world's sedimentary cover. Where two fluid bodies of different density are mixed together, the less dense fluid will tend to move above the denser one. Conversely, the denser fluid will tend to flow downwards. **Aqueous** density flows may be caused by differences of temperature, salinity and suspended sediments. **Glacial** melt streams and certain polar currents tend to flow under gravity beneath warmer, less-dense water bodies. Water discharged by rivers in temperate latitudes often flows out for considerable distances from the shore above the denser, more saline sea-water.

Eolian processes involve both traction carpets and suspensions (dust clouds). Turbidity flows are essentially unknown except in volcanic gas clouds termed "nuee ardentes". These are masses of hot volcanic gases with suspended ash and glass shards. These masses move down the sides of volcanoes at great speed. The resultant deposit, termed an ignimbrite, may have been formed at sufficiently high temperature for the ash particles to be welded together. Modern eolian sediment transport and deposition occurs in three situations. They are found in the arid desert areas of the world, such as the **Sahara**. They are found erratically developed around ice caps, where the climate may have considerable **precipitation**, but this is often seasonal and ice-bound. Dunes also occur on the crests of barrier islands and beaches in diverse climates.

Studies have shown how sediment, blown by the wind, moves by sliding and saltation just like particles in water. Quartz particles of about 0.10mm (very fine sand) are first to move in a rising wind. Silt and clay need velocities as strong as those for fine-medium sand to initiate movement. **Ripples**, dunes and plane beds are all common eolian sand bed forms. It is a matter of observation that ripples are blown out on both dunes and plane beds during sandstorms, to be rebuilt as the wind vanes.

Several types of sedimentary deposit are associated with glaciation. Ice itself transports and deposits one rock type only, termed diamictite. This is a poorly-sorted sediment ranging from **boulders** down to clay grade. Much of the clay material is composed of diverse minerals, but largely silica, formed by glacial pulverization. Ice, formed from compacted snow, moves both in response to gravity, in valley glaciers, and in response to horizontal pressures in continental ice sheets. Ice movement is very slow, compared with aqueous or eolian flow velocities. On the other hand it is highly erosive, breaking off boulders from the rocks over which it moves. The detritus which is caught up in the base of a glacier is transported in the direction of ice flow. It is unaffected by the sorting action found in eolian and aqueous action.

New Words and Expressions

1. weathering *n.* 风化作用
2. aqueous *adj.* 水的，水成的
3. traction *n.* 牵引作用
4. bed load 底负载
5. winnow *vt.* 扬，簸，筛选
6. suspension *n.* 悬浮液
7. unidirectional *adj.* 单向的
8. estuary *n.* （江河入海的）河口
9. turbidity *n.* 浊流
10. interbedded *adj.* 夹层之间的
11. eolian *adj.* 风的，风成的
12. Sahara *n.* 撒哈拉沙漠
13. precipitation *n.* 某地区降雨的量
14. ripple *n.* 涟漪，波痕
15. alluvium *n.* 冲积层
16. playa *n.* 干盐湖，干荒盆地
17. glacial *adj.* 冰川的
18. boulder *n.* 砾石，巨砾

Lesson 10 ▸ Diagenesis

Diagenesis are the chemical and physical changes that sediments undergo during and after their accumulation, but before **consolidation**. Diagenetic pathways depend on the initial sediment composition and grain size, depositional environment, temperature and pressure conditions during progressive burial, and depth of burial. Diagenetic processes modify porosity and permeability largely. In general, constructive diagenesis lowers porosity and permeability of sandstones or carbonate rocks, destructive diagenesis creates new porosity and permeability.

Diagenetic studies become increasingly important to oil and gas exploration. Wilson (1977) even proposed a third category of hydrocarbon traps besides structural and stratigraphic traps, namely the diagenetic trap. The ultimate goal of diagenetic studies in petroleum exploration is to predict type and magnitude of diagenetic alterations to be found in subsurface target areas prior to leasing and/or drilling. Thus, diagenetic studies will enhance better predictions of new plays.

All sedimentary rocks are more or less subject to diagenesis. In the frame of petroleum sedimentology one focuses on the diagenesis of source rocks, **claystones** and **shales**, and of reservoir rocks, namely sandstones and carbonate rocks. Included is also the diagenesis of organic substances, oils, **brines**, and pore waters.

Porosity and **permeability** are the most important parameters in oil and gas exploration and production which are influenced by type and intensity of diagenesis.

Compaction is the mechanical process responsible for the loss of porosity and permeability in sedimentary rocks after deposition. During compaction grains move closer together by rotation and slippage of grains, by flexible, ductile, and brittle grain deformation, and by pressure solution. The porosity of sand can be reduced from 40% initial porosity to 28% by sole grain rearrangement.

Beside the compactional effects, porosities of shales, sandstones, and carbonate rocks decrease normally with depth of burial. Thus, porosities of sands range from an initial porosity of 45% to practically zero for compacted and cemented sandstones. However, even in tight sandstones an **irreducible** porosity of approximately 2% remains. The porosity - depth interrelation, an exploration parameter of utmost significance, differs from borehole to borehole and basin to basin depending upon sandstone composition, geothermal gradient, geotectonic history, and local factors.

Primary factors which influence considerably the course of diagenesis in sandstones and carbonate rocks are the initial mineral composition of the **detritus** and the initial composition of pore solutions. The mineral composition of detrital grains, i. e. the mechanical and chemical environment of the detrital mineral assemblage, may cause differences in compaction. A rigid sandstone rich in quartz and **chert clasts** shows less compaction than a soft sandstone rich in shale clasts. The physical chemistry of the environment of deposition, especially the chemical environment of pore solutions, may cause **cryptocrystalline ankerite**, **dolomite**, or siderite to be formed soon after deposition, especially in siltstones and sandstones of swamp and organic-rich **lagoonal** environments.

Main motor of diagenesis are the changing chemical and physical conditions, mainly associated with the increase in temperature and pressure and/or with augmenting depth of burial.

Temperature plays a major role in initiating and controlling diagenetic reactions. The average geothermal gradient is about 3.3℃/100m, but it may vary from region to region. Higher geothermal gradients occur in sedimentary sequences surrounding igneous intrusions where short-distance gradients of several hundred degrees/100m are recorded. In regions of a high geothermal gradient porosities are normally lower than in regions with a low geothermal gradient.

Diagenesis of organic matter is influenced mainly by biological activity at an early stage as well as by temperature and pressure at a later stage. During early diagenesis, microbial activity is the main agent of organic matter transformation. The depth interval of early diagenesis ranges from zero down to 2,000m. Temperature and pressure increase slightly. Aerobic microorganisms in the uppermost sediment layer consume free oxygen. Anaerobes reduce sulfates to sulfides. The decomposing organic matter is converted into carbon dioxide, **ammonia**, and water. Simultaneously the Eh decreases abruptly and the pH increases slightly. Biogenic $CaCO_3$ and SiO_2 dissolve and normally re-precipitate together with **authigenic sulfides** and **siderite**. With advancing diagenesis, the organic matter grades mainly into **kerogen**. When large amounts of organic matter derived from plants are deposited, peat and subsequently brown coals (lignite and subbituminous coal) are formed. **Methane** is the most important hydrocarbon generated during diagenesis. Additionally, decomposing organic matter releases CO_2, H_2O, and some heavy heteroatomic compounds late in diagenesis.

Diagenetic processes affecting sands after deposition modify initial porosity, permeability, entry pressure, and irreducible **water saturation**. The most important of theses processes is cementation apart from grain fracturing and pressure solution. Mineral cementation normally reduces pore dimension and permeability, and increases entry pressure because of enlarging the internal surface area. Introduction of diagenetic clay minerals can also negatively affect entry pressure, permeability, and irreducible water saturation without changing porosity appreciably. This effect may be caused by modifications of pore size, crystal shape, and surface chemistry of different clay minerals.

Diagenetic processes act on original lime muds and consolidated limestones including: organic decay, **neomorphism**, **cementation**, solution - welding, telogenetic **dissolution**, carbonate-grain rearrangement, **pressure solution**, **stylolitization**, and **fracturing**.

Organic decay causes breakdown and grain-size reduction of biogenic **calcite**. Organic rims encrust particles and burrowing loosens the sediment. Neomorphism converts unstable marine aragonite and high-Mg calcite into low-Mg calcite or dolomite through submicroscopic replacement processes. Cementation destroys initial porosity within and between the coarser carbonate rock particles.

Solution-welding takes place in lime sands and lime muds. Contacts of skeletal and non-skeletal grains are modified by dissolution under the influence of freshwater near the surface or at shallow depths of burial. Solution-welding is thought to be involved in the fabric transformation from lime mud to microcrystalline calcite such as in partly cemented chalk.

Telogenetic dissolution causes secondary porosity cutting across grains and cements and enlarging **intergranular pores**. Even major caverns may form along joints and fractures.

Carbonate-grain rearrangement prior to burial commonly forces grains into planar contact early in diagenesis. Pressure solution, stylolitization, fracturing, and vein formation are late diagenetic processes subsequent to **lithification**.

New Words and Phrases

1. diagenesis *n.* 成岩作用
2. consolidation *n.* 固结(作用)
3. porosity *n.* 孔隙度
4. permeability *n.* 渗透性
5. claystone *n.* 黏土岩
6. shale *n.* 页岩
7. brine *n.* 卤水, 海水
8. irreducible *adj.* 不可还原的
9. detritus *n.* (侵蚀形成的)岩屑
10. chert clasts 硅质碎屑
11. cryptocrystalline *n.* 隐晶质
12. ankerite *n.* 铁白云石
13. dolomite *n.* 白云石
14. lagoonal *adj.* 潟湖的
15. ammonia *n.* 氨
16. authigenic *adj.* 自生的
17. sulfide *n.* 硫化物
18. siderite *n.* 菱铁矿
19. kerogen *n.* 油母岩质,干酪根
20. methane *n.* [化]甲烷,沼气
21. water saturation 饱和度
22. neomorphism *n.* 新生变形作用
23. cementation *n.* 胶结
24. dissolution *n.* 分散, 溶解
25. pressure solution 压溶
26. stylolitization 新生变形作用
27. fracture *n.* 破裂, 碎裂,龟裂
28. calcite *n.* 方解石
29. intergranular pores 晶粒间的
30. lithification *n.* 岩化

Lesson 11 ▸ Clastic Rocks

Clastic sedimentary rocks are rocks composed predominantly of broken pieces or clasts of older weathered and eroded rocks.

Composition

The composition of siliciclastic sedimentary rocks includes the chemical and mineralogical components of the framework as well as the cementing material that make up these rocks. There are four categories: major minerals, accessory minerals, **rock fragments**, and chemical cements.

Major minerals can be categorized into subdivisions based on their resistance to chemical decomposition. Those that possess a great resistance to decomposition are categorized as stable, while those that do not are considered less stable. The most common stable mineral in siliciclastic sedimentary rocks is quartz. **Quartz** makes up approximately 65 percent of framework grains present in sandstones and about 30 percent of minerals in the average shale. Less stable minerals present in this type of rocks are feldspars, including both potassium and plagioclase feldspars. **Feldspars** comprise a considerably lesser portion of framework grains and minerals. They only make up about 15 percent of framework grains in sandstones and 5% of minerals in shales. Clay mineral groups are mostly present in mudrocks (comprising more than 60% of the minerals) but can be found in other siliciclastic sedimentary rocks at considerably lower levels.

Accessory minerals are associated with those whose presence in the rock are not directly important to the classification of the specimen. These generally occur in smaller amounts in comparison to the quartz, and feldspars. Furthermore, those that do occur are generally **heavy minerals** or coarse grained micas (both **Muscovite** and **Biotite**).

Rock fragments also occur in the composition of siliciclastic sedimentary rocks and are responsible for about 10~15 percent of the composition of sandstone. They generally make up most of the gravel size particles in conglomerates but contribute only a very small amount to the composition of mudrocks. Though rock fragments are not always sedimentary rocks in origin. They can also be **metamorphic** or **igneous**.

Chemical cements vary in abundance but are predominately found in sandstones. The two major types, are silicate based and carbonate based. The majority of silica **cements** are composed of quartz but can include **chert**, **opal**, feldspars and **zeolites**.

Classification

Siliciclastic sedimentary rocks are composed of mainly silicate particles derived by the weathering of older rocks. siliciclastic sedimentary rocks are classified according to grain size into three major categories; **conglomerates**, sandstones, and mudstones (Table 11. 1). The term clay is used to classify particles smaller than 0. 0039mm. However, term can also be used to refer to a family of sheet silicate minerals. Silt refers to particles that have a diameter between 0. 062 and 0. 0039mm. The term mud is used to refer to when clay and silt particles often mix to create aggregates sediments. Hence, the term mudrock is used to refer to rocks that are composed of silt and clay particles. Furthermore, particles that reach diameters between 0. 062mm and 2mm fall into the category of sand. When sand is cemented together and lithified it becomes known as sandstone. Any particle that is larger than 2mm is considered gravel. This category includes pebbles, cobbles and boulders. Like sandstone, when gravels are lithified they are considered conglomerates.

Table 11. 1 Classification of clastic rocks

<table>
<tr><th>Particle Diameter(mm)</th><th colspan="2">Particle Name</th><th colspan="2">Rock Name</th></tr>
<tr><td>>256</td><td>Boulder</td><td rowspan="3">Gravel</td><td colspan="2" rowspan="3">Conglomerate/Breccia</td></tr>
<tr><td>64~256</td><td>Cobble</td></tr>
<tr><td>2~64</td><td>Pebble</td></tr>
<tr><td>0. 062~2</td><td colspan="2">Sand</td><td colspan="2">Sandstone</td></tr>
<tr><td>0. 0039~0. 062</td><td>Slit</td><td rowspan="2">Mud</td><td>Siltstone</td><td rowspan="2">Mudstone</td></tr>
<tr><td><0. 0039</td><td>Clay</td><td>Claystone/Shale</td></tr>
</table>

Conglomerates

Conglomerates are coarse grained rocks dominantly composed of gravel sized particles that are typically held together by a finer grained martix. These rocks are often subdivided up into conglomerates and breccias. The major characteristic that divides these two categories is the amount of rounding. The gravel sized particles that make up conglomerates are well rounded while in **breccias** they are angular. Conglomerates are common in stratigraphic successions of most, if not all ages but only make up one percent or less, by weight of the total sedimentary rock mass.

Sandstones

Sandstones are medium grained rocks composed of rounded or angular fragments of sand size, that often but not always have a cement uniting them together. These sand size particles are often quartz minerals but there are a few common categories and a wide variety of classification schemes that are given are used to classify sandstones based on composition.

Mudstones

Rocks that are classified as mudstones are very fine grained. Silt and clay represent at least

50% of the material that mudstones are composed of. Classification schemes for mudstones tend to vary but most are based on the grain size of the major constituents. In mudstones, these are generally silt, and clay. Mudstones that are composed mainly of silt particles are classified as siltstones. In turn, rocks that possess clay as the majority particle are called claystones. In geology, a mixture of both silt and clay is called mud. Rocks that possess large amounts of both clay and silt are called mudstones. In some cases the term shale is also used to refer to mudstones and is still widely accepted by most. However, others have used to term shale to further divided mudstones based on the percentage of clay constituents. The plate like shape of clay allows its particles to stack up one on top of another creating laminae or beds. The more clay present in a given specimen, the more laminated a rock is. Shale, in this case, is reserved for mudstones that are laminated, while mudstone refers those that are not.

New Words and Phrases

1. quartz *n.* 石英
2. feldspar *n.* 长石
3. rock fragment 岩屑
4. heavy minerals 重矿物
5. biotite *n.* 黑云母
6. muscovite *n.* 白云母
7. cement *n.* 胶结物
8. chert *n.* 玉髓
9. opal *n.* 蛋白石
10. zeolite *n.* 沸石
11. conglomerate *n.* 砾岩
12. boulder *n.* 巨砾
13. cobble *n.* 粗粒
14. pebble *n.* 细砾
15. sand *n.* 砂
16. silt *n.* 粉砂
17. clay *n.* 黏土
18. metamorphic *adj.* 变质的
19. igneous *adj.* 火成的
20. breccia *n.* 角砾岩
21. sandstone *n.* 砂岩
22. siltstone *n.* 粉砂岩
23. mudstone *n.* 泥岩
24. shale *n.* 页岩

Lesson 12 ▸ Sandstones

Sandstone is a sedimentary rock and one of the most common types of sedimentary rock and is found in sedimentary basins throughout the world. It is composed of sand-size (1/16~2mm) grains rock fragment, mineral and organic material. Sand-size particles range in size from millimeters in diameter. Also the cementing material binds the sand grains together and may contain a matrix of silt-or clay-size particles that occupy the spaces between the sand grains.

Sandstones are significant for a variety of reasons. Volumetrically they constitute between 10 and 20 percent of the Earth's sedimentary rock record. They are resistant to erosion and therefore greatly influence the landscape. Sandstones are economically important as major reservoirs for both petroleum and water, as building materials, and as valuable sources of metallic ores. Most significantly, they are the single most useful sedimentary rock type for deciphering Earth history. Sandstone mineralogy is the best indicator of sedimentary provenance: the nature of a sedimentary rock source area, its composition, relief, and location. Sandstone textures and sedimentary structures are also reliable indexes of the transportational agents and depositional setting.

Sandstone Composition

Chemical Composition of sandstone usually quartz framework grains are the dominant mineral in clastic sedimentary rocks. Because of the exceptional physical properties such as hardness and chemical stability, physical properties of these quartz grains survive multiple recycling events and also allowing the grains to display some degree of rounding. Quartz grains evolve from plutonic rock, which are felsic in origin and also from older sandstones that have been recycled.

Second most abundant mineral is feldspathic framework grains. Feldspar can be separate into two subdivision. They are **alkali feldspars** and **plagioclase feldspars**. Feldspars minerals is distinguished under a petrographic microscope. Alkali feldspar is a group of minerals in which the chemical composition of the mineral can range from $KAlSi_3O_8$ to $NaAlSi_3O_8$, this represents a complete solid solution. Plagioclase feldspar is a complex group of solid solution minerals that range in composition from $NaAlSi_3O_8$ to $CaAl_2Si_2O_8$.

Lithic framework grains are pieces of ancient source rock that have yet to weather away to individual mineral grains, called lithic fragments or clasts. Lithic fragments can be any fine-grained or coarse-grained igneous, metamorphic, or sedimentary rock, although the most common lithic fragments found in sedimentary rocks are clasts of volcanic rocks.

Accessory minerals are small percentage of the grain in a sandstone. Common accessory minerals include micas (muscovite and biotite), **olivine**, **pyroxene**, and **corundum**. Many of these accessory grains are denser than silicates minerals in the rocks. These heavy minerals are more durability to weathering and can be used as an indicator of sandstone maturity through the ZTR index.

Common heavy minerals include **zircon**, tourmaline, **rutile** (hence ZTR), **garnet**, magnetite, or other dense, resistant minerals derived from the source rock.

Matrix is present within fractured pore space between the framework grains. This pore space can be separate into the two classes. They are **Arenites** and **Wackes**. Arenites are texturally clean sandstones that are free of or have very little matrix. Wackes are texturally dirty sandstones that have a significant amount of matrix.

Cement binds the siliciclastic framework grains together. Cement is secondary deposition minerals formed after burial of sandstone. These cementing materials may be either silicate minerals or non-silicate minerals, such as **calcite**. Silica cement can consist of either quartz or opal minerals. Calcite cement is the most common carbonate cement. Calcite cement is an assortment of smaller calcite crystals. Other minerals that act as cements include **hematite**, **limonite**, feldspars, **anhydrite**, **gypsum**, **barite**, **clay minerals**, and zeolite minerals.

Classification of Sandstones

Texture and mineralogical properties are used for sandstone classification, though considerable debate exists as to which properties to emphasize. The two major classes of sandstone are arenite and wacke (Figure 12.1). The boundary between the two is based on the amount of matrix present in the sample. Arenites contain less matrix than wacke. Though the exact boundary is debated, often 5 percent matrix is the accepted value, whereas some experts place this boundary at 15 percent. Another common classification uses the name arkose for rocks rich in feldspar. These names are further modified by the components present in the rock. **Quartz arenites**, **feldspathic arenites**, and **lithic arenites** are common examples.

Quartz arenites are composed of more than 90 percent of siliceous grains (quartz, chert, other quartzose rocks fragments), generally well lithified and well cemented, and cemented with silica or carbonate cement. It is associated in stable cratonic environments such as eolian, beach and shelf environments. Fossil preservation is poor, but trace fossils such as burrows of the skolithos facies which occurring in intertidal zones may be abundant in some shallow marine quartz arenites. Quartz arenites are first-cycle deposits derived from primary crystalline or metamorphic rocks which had a weathering process. Quartz arenites are common in the geologic record. They are commonly white or light, may be stained red, pink, yellow and brown by iron oxides.

Feldspathic arenites have pink or red colours because of the abundant presence of

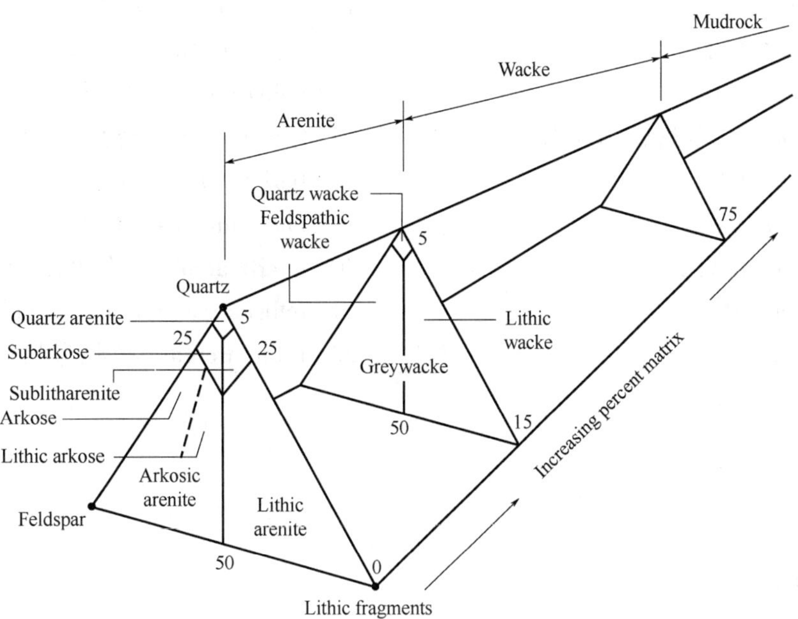

Figure 12. 1 Classification of sandstones

potassium feldspars or iron oxides, may be light gray or white. They contain less than 90 percent quartz, more feldspars than lithic fragments, with minor amount of micas and heavy minerals. They typically have medium to coarse grains and may contain high percentages of subangular to angular grains, contain more than 15 percent of matrix and poorly–well sorted of grain sorting. It means that commonly texture are immature–submature. It typically occurred in cratonic or stable shelf settings, and less typically deposited in unstable basins or other deeper water.

Lithic arenites are characterized by unstable rock fragment such as volcanic, metamorphic clast and other, also chert or quartzose rocks can be fragments of lithic arenites. With less than 90 percent quartzose grains and more rock fragments, lithic arenites have gray, salt-and-pepper to uniform medium and gray. Lithic arenites tend to have amount of matrix, of course it can be called lithic wackes which has poorly to very poorly sorted, but it can be well sorted. Most lithic sandstone are texturally immature to submature, occur in association with fluvial conglomerates and other fluvial deposits, or in association with deeper marine conglomerates, shale, cherts, and submarine basalts.

New Words and Phrases

1. alkali feldspars 碱性长石
2. plagioclase feldspar 斜长石
3. olivine *n.* 橄榄石
4. pyroxene *n.* 辉石

5. corundum *n.* 刚玉
6. rutile *n.* 金红石
7. zircon *n.* 锆石
8. chlorite *n.* 亚氯酸盐,绿泥石
9. arenites *n.* 砂岩
10. wackes *n.* 杂砂岩
11. garnet *n.* 石榴子石
12. matrix *n.* 杂基
13. calcite *n.* 方解石
14. hematite *n.* 赤铁矿
15. limonite *n.* 褐铁矿
16. anhydrite *n.* 硬石膏
17. gypsum *n.* 石膏
18. barite *n.* 重晶石
19. clay minerals 黏土矿物
20. quartz arenite 石英砂岩
21. feldspathic arenite 长石砂岩
22. lithic arenite 砂屑砂岩

Lesson 13 ▸ Mud and Mudstones

Detrital fine-grained sediments, all those composed of 50 percent or more of particles smaller than 62 micrometers, include terrigenous mud and mudstone, carbonate mud (**micrite**) and mudstone and silt and siltstone plus deep sea ooze and **sapropel**. All of these may locally grade into one another and into carbonates and sandstones. Here we consider primarily the **terrigenous** mud-rich deposits and to a lesser degree their commonly associated silts and siltstones, which are nearly always present. We use the term mud for an unconsolidated deposit and the term mudstone for its lithified equivalent.

Mudstones provide us with many insights into major processes of sedimentation and thus effectively complement studies of both sandstones and carbonates. Mudstones, for example, help us identify sea level changes, provide key marker beds for both local correlation in **reservoirs** and, because of their great lateral continuity, for regional correlations over hundreds of kilometers. And it is the mudstones that, better than any other **lithology**, permit us to identify the paleo oxygen levels of ancient basins and explore their low-energy transport processes. In addition, most **Mesozoic** and **Cenozoic** shales are rich in pelagic faunas, which yield refined global age determinations. The presence and abundance of mudstones is a key factor for limiting carbonate and chemical deposition. In addition, mud and mudstone are linked to all the other fine-grained rocks. Muds and mudstones are commonly interbedded with silts and siltstones and grade into carbonate mud, marl and also sandstone, limestone and **dolostone**. Fine-grained **volcanic debris** is also common along active margins. Thus the study of mudstones connects us to most sedimentary rocks and has rewards totally different from the terrigenous sandstones, which provide us with "big picture **tectonics**" and the distribution of high-energy environments in ancient basins.

The study of mudstones is clearly not only the natural complement to the study of sandstones and carbonates but, in addition, mudstones are principal **source rocks** for petroleum and for the mineralized fluids that source many sedimentary ores. And because of their low permeability, muds and mudstones inhibit the vertical flow of petroleum, mineralized fluids and fresh water. Thus the distribution and continuity of mudstones in a reservoir is all important for its development, because mudstones segregate the reservoir into separate, independent or semi-independent parts. And finally, abundance also shows the importance of the study of muds and mudstones: they are by far the dominant sedimentary rock. Rare, **fine-grained rocks** that either grade into or are closely associated with mudstones include phosphorites and some **iron** stones and ores as well as the more common carbonate muds, marls and **volcanic ashes**.

Important controls on mud include its supply, which relate to the climate and relief of the source region; low turbulence in the **water column**, especially near or at the bottom; the oxygen level at the bottom, which governs total organics, color, lamination and faunal activity; and the rate of sedimentation. The slow settling velocity of fine particles ensures far-ranging dispersion by even weak currents and explains the wide extent of many mudstones, even the thinnest of beds.

Changes in relative sea level are a major control on the deposition of marine mud and depend on global **eustatic changes** in sea level, subsidence rates of basins and the **influx** of mud. The sum of these three independent factors determines the space available for sediment accumulation, called **accommodation**. The greater this space, the deeper the water and the more likely it is that mud will accumulate. On **continental shelves** and **cratons**, supply permitting, changes in relative sea level shift mud deposition from far inshore during high stands to beyond the **shelf edge** during low stands of the sea. In this scenario, the low energy environments of mud deposition shift back and forth from shallow to deep water, from well to poorly oxygenated water. They are deposited with widely different sedimentation rates and thus have different colors, textures, structures, and faunal and facies associations. On cratons lacking shelf breaks, a relative rise of sea level tends to produce vast, thin sheets of shale, whereas on a steep shelf or **ramp** there is little migration, geometry is wedge like, and **resedimentation** downdip is important.

Thick sections of mudstones accumulate in rapidly subsiding basins supplied either by large rivers draining to passive margins or by many small rivers closely coupled to actively subsiding foreland, forearc, backarc, and pull-apart basins. On cratons and on **passive margins** lacking large rivers, thinner but widely traceable shales commonly are linked to distal **deltas**, as they are large **submarine fans** offshore.

In all of these basins, oxygen levels at the sediment water interface and in the mud control not only bottom fauna, but also the color, **pyrite** abundance, and organic content of the mud plus its degree of **bioturbation**-a key factor for the preservation of lamination. Thus black shales typically are well laminated, have a sparse bottom fauna with low diversity and high total organic carbon and pyrite, whereas the common, greenish-gray mudstone has little pyrite and is likely to be bioturbated and poorly laminated. A secondary control on many of the above characteristics is the rate of sedimentation-how long the mud remains at or near the sediment-water interface.

Petrographically, most muds and mudstones consist of a mixture of metastable clay minerals and much less reactive silt-sized quartz and feldspar debris. The latter occurs either as millimetric wisps and laminations or as scattered silt grains plus fine grained organic debris of many different kinds-a good example of hydraulic equivalence depositing many unlike particles together. In many mudstones, the proportion of terrigenous silt may equal or exceed the proportion of clays. The proportions of silt and organic material greatly influence the physical properties of mud and mudstone, but the dominant control on the strength of a mud is still its water content.

Mud deforms easily both on land and underwater and flows at exceedingly low angles. After burial, rapidly deposited mud can become over pressured and form spectacular diapirs rising hundreds of meters into and through a sedimentary pile. Low angle slides and mud lumps are especially common in low energy, rapidly deposited deltas rich in mud. Mudstones, because of their low shear strength in comparison to sandstones and carbonates, deform and fail easily. These commonly form the soles of low angle over thrusts and, at the surface, are likely to fail, where they dip toward a valley. Artificial cuts in muds and shales are notably unstable and require special engineering attention even where dips are very shallow.

The **expulsion of water** from mud as it is compacted is central to its role in ancient basins, because it concentrates petroleum in traps and leads to the deposition of many sedimentary ores. With burial, the density of shale increases, rising temperatures and changing pore water chemistry produce many mineralogical changes. Transformations include glass to zeolites, clays to feldspar, **smectite** to **illite**, **kaolinite** to **chlorite**, opal A to opal CT and quartz, and complex changes in the organic matter (kerogen). New minerals include pyrite, marcasite and cements of iron, carbonate and phosphate.

New words and Phrases

1. micrite *n.* 泥晶方解石
2. sapropel *n.* 腐殖泥
3. terrigenous *adj.* 陆源沉积的
4. reservoir *n.* 储层
5. lithology *n.* 岩性
6. Mesozoic *n.* 中生代
7. Cenozoic *n.* 新生代
8. dolostone *n.* 白云岩
9. volcanic debris 火山碎屑
10. tectonic *adj.* 大地构造的
11. source rock 烃源岩
12. fine-grained rocks 细粒岩
13. iron *n.* 铁矿(铁质)
14. volcanic ash 火山灰
15. water column 水柱
16. eustatic change 海平面升降变化
17. influx *n.* 流入
18. accommodation *n.* 可容纳空间
19. continental shelf 大陆架
20. craton *n.* 克拉通
21. shelf edge 陆台外缘
22. ramp *n.* 斜坡
23. resedimentation 再沉积作用
24. passive margin 被动陆缘
25. delta *n.* 三角洲
26. submarine fan 海底扇
27. pyrite *n.* 黄铁矿
28. bioturbation *n.* 生物扰动作用
29. expulsion of water 排水作用
30. smectite *n.* 蒙皂石
31. illite *n.* 伊利石
32. kaolinite *n.* 高岭石
33. chlorite *n.* 绿泥石

Lesson 14 ▶ Carbonate Rocks

Carbonate rocks are the most abundant nonterrigenous sedimentary rocks, constituting roughly one-tenth of the Earth's sedimentary shell.

The carbonate rocks are extremely complex in their **genesis**, diagenesis, and petrophysics. There are a number of reasons for this. Carbonate rocks are intrabasinal in origin. Unlike terrigenous sediments they are easily weathered and their weathering products are transported as solutes. Carbonate rocks are, therefore, deposited at or close to their point of origin. Most carbonate rocks are organic in origin. They contain a wide spectrum of particle sizes, ranging from whole shells to lime mud of diverse origin. These sediments are deposits with a high **primary porosity**. However, the carbonate minerals are chemically unstable. This combination of high primary porosity and **permeability**, coupled with chemical instability, is responsible for the complicated diagenesis of carbonate rocks, and hence for the problems of locating **aquifers** and hydrocarbon reservoirs within them.

The following brief account of carbonates first defines their mineralogy, then describes their **petrography** and classification.

Carbonate Minerals

It is necessary to be familiar with the common carbonate minerals to understand the complex diagenetic changes of carbonate rocks. **Calcium carbonate** ($CaCO_3$) is the dominant constituent of modern carbonates and ancient limestones. It occurs as two minerals, **aragonite** and **calcite**. Aragonite crystallizes in the **orthorhombic crystal system**, while calcite is **rhombohedral**. Calcite forms an **isomorphous series** with **magnesite** ($MgCO_3$).

Dolomite is another important carbonate mineral, giving its name also to the rock. Dolomite is **calcium magnesium carbonate** ($CaMg(CO_3)_2$). Isomorphous substitution of some magnesium for iron is found in the mineral termed ferroan dolomite or ankerite $Ca(MgFe)(CO_3)_2$. Unlike calcite and aragonite, dolomite does not originate as skeletal material. Dolomite is generally found either crystalline, as an obvious **secondary replacement** of other carbonates, or as a primary or **penecontemporaneous** replacement mineral in **cryptocrystalline** form. **Siderite**, iron carbonate ($FeCO_3$), is one of the rarer carbonate minerals. It occurs, apparently as a primary precipitate, in **ooliths**. These "spherosiderites", as they are termed, are found in rare restricted marine and freshwater environments. Siderite forms diagenetically during early burial while the host sediment is still uncompacted. Its formation is favored by alkaline reducing condition.

Physical Components of Carbonate Rocks

Carbonate rocks, like sandstones, have four main components: grains, **matrix**, **cement**, and pores (Table 14. 1). Unlike sandstones, however, the grains of carbonate rocks, though commonly monomineralic, are textually diverse and polygenetic.

Table 14. 1 Main Components of Carbonate Rock

Main components	Sub-components	Categories of sub-components
Ⅰ. Grains	(a) Detrital grains	Lithoclasts
		Intraclasts
	(b) Skeletal grains	
	(c) Peloids (including fecal pellets)	
	(d) Lumps	Composite grains
		Algal lumps
	(e) Coated grains	Ooliths
		Pisoliths
		Oncolites
Ⅱ. Matrix		Micrite
		Clay
Ⅲ. Cement		
Ⅳ. Pores		

Grains are the particles that support the framework of a sediment. They are thus generally sand grade or larger. Carbonate grains are of many types. First are the detrital grains, which are of two types. They include intraclasts, or lithoclasts. These are grains of noncarbonated material which originated outside the depositional basin. Quartz grains are a typical example of lithoclasts and, as they increase in abundance, limestones grade into sandy limestones and thence to **calcareous sandstones**. The second type of detrital grains are intraclasts. These are fragments of reworked carbonate rock which originated within the depositional basin.

The most important of all grain types is skeletal grains, individual grains of which are termed bioclasts. They are composed of aragonite or calcite with varying amounts of trace elements. The size of skeletal particles is naturally very variable, ranging down from the largest shell to individual disaggregated microscopic crystals. **Peloids** are a third major grain type. These were first defined as structureless cryptocrystalline carbonate grains of some 20~60μm in diameter. Lumps are another important carbonate grain type. These are **botryoidal grains** which are composed of several peloids held together. They are sometimes termed "composite grains" or grapestone. Last of the grain types to consider are the coated grains. These are grains that show a concentric or radial arrangement of crystals about a nucleus. The most common coated grains are ooids. These are rounded grains of medium to fine grain size which generally occur gregariously in sediments termed oolites, devoid of other grain types or matrix. **Pisoids** are coated grains several millimeters in diameter. They form in caverns (cave pearls). A rock composed of pisoids is termed a "pisolith". **Oncoids** are a third type of coated grain. These are several centimeters in diameter and irregularly shaped. In contrast to ooids, pisoids and oncoids are indicators of low-energy environments. A rock composed of oncoids is termed an "oncolite".

Carbonate mud is termed **micrite**. The upper size limit of micrite is variously taken as 0.03~0.04 mm in diameter. Micrite may be present in small quantities as a matrix within a grain-supported carbonate sand, or may be so abundant that it forms a carbonate mudrock, termed "**micrite**" or "**calcilutite**". Most modern lime muds are composed of aragonite; their lithified fossil analogs are made of calcite. Micrite can also form as a cryptocrystalline cement in certain circumstances. Because of this it must be used with care as an index of depositional energy.

The third component of carbonate rocks is cement. By definition this applies to crystalline material that grows within the sediment fabric during diagenesis. The most common cement in limestones is calcite, termed "spar" or **sparite**. Other cements in carbonate rocks include dolomite, **anhydrite**, and silica.

It is the custom to restrict the term "cement" to the growth of crystals within a pore space. This has also been termed "drusy crystallization". This type of spar is genetically distinguishable from neomorphic spar which grows by replacement of preexisting carbonate. Recrystallization is the development of calcite spar by the enlargement of preexisting calcite crystals.

New Words and Phrases

1. genesis *n.* 成因，起源
2. permeability *n.* 渗透性，渗透率
3. aquifer *n.* 蓄水层，含水层
4. petrography *n.* 岩相学，岩类学
5. aragonite *n.* 文石
6. calcite *n.* 方解石
7. rhombohedral *adj.* 斜方六面体的，菱面体的
8. magnesite *n.* 菱镁矿
9. dolomite *n.* 白云石，白云岩
10. cryptocrystalline *adj.* 潜晶质的，隐晶质的
11. penecontemporaneous *adj.* 准同期的，准同生的
12. siderite *n.* 菱铁矿
13. oolith *n.* 鲕粒
14. matrix *n.* 基质
15. cement *n.* 胶结物
16. peloid *n.* 球粒
17. pisoid *n.* 豆粒
18. oncoid *n.* 似核形石
19. micrite *n.* 泥晶，泥晶灰岩
20. calcilutite *n.* 泥屑灰岩
21. sparite *n.* 亮晶，亮晶方解石
22. anhydrite *n.* 硬石膏，无水石膏
23. primary porosity 原生孔隙度
24. calcium carbonate 碳酸钙
25. orthorhombic crystal system 斜方晶系
26. isomorphous series 类质同象系列
27. calcium magnesium carbonate 碳酸钙镁
28. secondary replacement 次生交代
29. calcareous sandstone 钙质砂岩
30. botryoidal grain 葡萄状颗粒

Lesson 15 ▸ Sedimentary Environments and Facies

The sedimentary environment is the specific depositional setting of a particular sedimentary rock and is unique in terms of physical, chemical, and biological characteristics. The physical features of a sedimentary environment include water depth and the velocity and persistence of currents. Chemical characteristics of an environment include the salinity (proportion of dissolved salts), acidity or basicity (pH), oxidation potential (Eh), pressure, and temperature. The biological characteristics are mainly the assemblage of fauna and flora that populate the setting. These conditions, combined with the nature of the transporting agent and the source area, largely determine the properties of the sediments deposited within the environment.

A number of ways of classifying depositional environments exist, but most modern schemes employ a geomorphologic approach. That is to say, an environment is defined in terms of a distinct geomorphic unit or landform, modern examples of which are readily visible for comparative purposes, e. g. , a river delta, an alluvial fan, a submarine fan, or the abyssal floor of an ocean basin. Individual environments are further grouped into three categories (Figure 15. 1):

(1) Continental environments;

(2) Transitional environments;

(3) Marine environments.

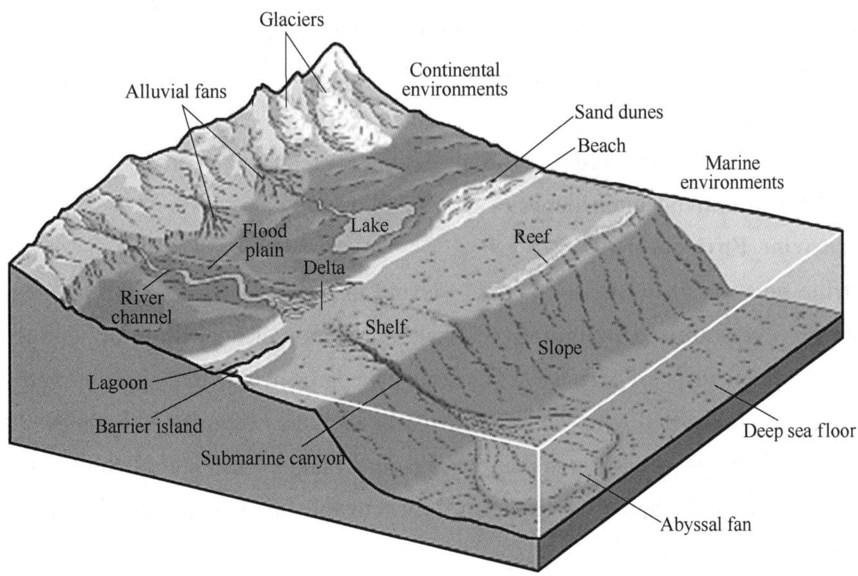

Figure 15. 1 Schematic diagram showing the main depositional environments

Continental Environments

The continental environments are those environments which are present on the terrestrial plains of continents. They are **fluvial** environments (rivers), **alluvial** fans, **lacustrine** environments (lakes), **aeolian** or eolian environments (deserts), paludal environments (swamps), and **glacial** environments.

1. Fluvial Environments

Fluvial environments are characterized by flow and deposition in river channels and associated overbank sedimentation. In the stratigraphic record, the channel fills are represented by lenticular to sheet-like bodies with scoured bases and channel margins, although these margins are not always seen. The deposits of gravelly braided rivers are characterized by cross-bedded conglomerate representing deposition on channel bars. Both sandy braided rivers and meandering river deposits typically consist of fining-upward successions from a sharp scoured base through beds of trough and planar cross-bedded, laminated and cross-laminated sandstone. Lateral accretion surfaces characterize meandering rivers that are also often associated with a relatively high proportion of overbank **facies**. Floodplain deposits are mainly alternating thin sandstone sheets and mudstones with palaeosols; small lenticular bodies of sandstone may represent crevasse splay deposition.

2. Alluvial Fans

Alluvial fan deposits are located near to the margins of sedimentary basins and are limited in lateral extent to a few kilometers from the margin. The facies are dominantly conglomerates, and may include matrix-supported fabrics deposited by debris flows, well-stratified gravels and sands deposited by sheet-flooded processes and in channels that migrate laterally across the fan surface. Alluvial and fluvial deposits will interfinger with lacustrine and/or aeolian facies, depending on the palaeoclimate, and many (but not all) river systems feed into marine environments via coasts, estuaries and deltas. Other characteristics of fluvial and alluvial facies include an absence of marine fauna, the presence of land plant fossils, trace fossils and palaeosol profiles in alluvial plain deposits.

3. Lacustrine Environments

Lacustrine deposits are very well sorted, devoid of coarse particles such as coarse sand or gravels, and are characterized by thin layers that reflect annual deposition of sediments. Lakes (lacustrine environments) are diverse. They may be large or small, shallow or deep, fresh water or salt water, and filled with terrigenous, carbonate, or evaporite sediments. Mudcracks, wave ripples, laminations, and varves may be present in lakes. The soil of the plain left behind may constitute fertile and productive farm land, due to the previous accumulation of lacustrine sediments (algal and bacterial).

4. Aeolian Environments

Aeolian sedimentary processes are those involving transport and deposition of material by the wind. The most obvious aeolian environments are the large sandy deserts in hot, dry areas of continents. Aeolian deposits occur mainly in arid environments where surface water is intermittent and there is little plant cover. They are less common outside of desert environments. Associated facies in arid regions are mud and evaporates deposited in ephemeral lakes and poorly sorted fluvial and alluvial fan deposits.

5. Paludal Environments

The swamps often form in low-lying areas in which parallel layered, organic-rich black shales and coal form. Swamps are areas of standing water with trees. Decaying plant matter accumulates to form peat, which may eventually become coal. The biological activity is frequent, so rocks of high organic content are abundant. The mixed muds & coal/peat/lignite, with some fluvial sediments are often associated with other environments (**deltaic**, fluvial, coastlines, etc.) and low energy. The swamp deposits plant-choked, periodically inundated environments.

6. Glacial Environments

Glacial environments are characterized by the dominance of ice as a geological agent. Glacial deposits are compositionally immature and tills are typically composed of detritus that simply represents broken up and powdered bedrock from beneath the glacier. Deposition of glacial material on land produces characteristic landforms and distinctive sediment characters, but these continental glacial deposits generally have a low preservation potential in the long term and are rarely incorporated into the stratigraphic record. Glacial processes which bring sediment into the marine environment generate deposits that have a much higher chance of long-term preservation.

Transitional Environments

The Transitional environments are those environments at or near where the land meets the sea. Transitional sedimentary environments include deltas, esturies, lagoons, beaches and barrier islands, salt marshes, and **tidal flats**.

1. Deltas

The deltas are large accumulations of sediments that are deposited where a river empties into a standing body of water. They are one of the most significant environments of sedimentation and include a number of sub environments such as stream channels, flood plain beaches, bars, and tidal flats. The deposit as a whole consists of a thick accumulation of sand, silt, and mud. The deltas are fan-shaped deposits of sediment, formed where a river flows into a standing body of water, such as a lake or sea. The coarser sediment (sand) tends to be deposited near the mouth of the river; finer sediment is carried seaward and deposited in deeper water. The alluvial-deltaic boundary occurs where the river begins to bifurcate and forms distributaries.

2. Esturies

The estuaries are the marine-influenced portion of drowned valleys. A drowned valley is the seaward portion of a river valley that becomes flooded with sea water when there is a relative rise in sea level. They are regions of mixing of fresh and seawater. Sediment supply to the estuary is from both river and marine sources, and the processes that transport and deposit this sediment are a combination of river and wave and/or tidal processes.

3. Lagoons

The lagoons are bodies of water on the landward side of barrier islands. They are protected from the pounding of the ocean waves by the barrier islands, reef, sandbank, and spit. The lagoon has low energy, so sediment is dominantly fine grained. The lagoons are coastal bodies of water that have very limited connection to the open ocean. The lagoons generally develop along coasts where there is a wave-formed barrier and are largely protected from the power of open ocean waves. A **lagoonal** succession is typically mudstone, with thin, wave-rippled sand beds.

4. Beaches and Barrier Islands

The beaches are the most recognizable transitional environments. The beaches form where wave energy washes silt and clay away, leaving larger sand particles behind. The silt and clay may then be deposited in other low energy transitional environments. They termed as tidal flats or in deep marine environments. The beaches and barrier islands are shoreline deposits exposed to wave energy and dominated by sand with marine fauna. The barrier islands are separated from the mainland by a lagoon. They are commonly associated with tidal flat deposits.

Marine Environments

The marine environments are in the seas or oceans. The marine environments include shallow marine environments, the continental shelf, slope, rise, and **abyssal** plain.

1. Shallow Marine Environments

Nearly flat, gentle sloping edge of the continent that extends under the ocean. The regions near the mouths of rivers are usually clastic dominated, because the critters that secrete $CaCO_3$ tend to have trouble living in muddy water. In closed basins, salinity levels of the water bodies may increase above those of seawater, as a result of excess evaporation concentrating the salts in solution. The sediments wash off of the continent onto the shelf. The size may range from coarse sand closest to the continent fining to clays. In some areas coral reefs or carbonate muds dominate the shelf.

2. Continental Shelf

The continental shelf is the flooded edge of the continent. It is relatively flat (with a slope of less than 0.1°), shallow (less than 200 m or 600 ft deep). It may be up to hundreds of miles wide. The continental shelves are exposed to waves, tides, and currents. They are covered by sand, silt, mud, and gravel. The flooding of the edges of the continents occurred when the glaciers melted at the end of the last Ice Age, about 10,000 years ago. The continental shelf is the shallow ocean surrounding the continent. The depth at the edge of the shelf is usually not more than 100 to 150 meters. Some sand and mud are carried to the edge of the continental shelf via **submarine canyons** which are like undersea river valleys.

3. Continental Slope and Continental Rise

The continental slope and continental rise are located seaward of the continental shelf. The continental slope is the steep (5° ~ 25°) "drop-off" at the edge of the continent. The continental slope passes seaward into the continental rise, which has a more gradual slope. The continental rise is at the base of the continental slope, where thick accumulations of sediment are deposited. The sediment builds up at the edge of the shelf. The resulting deposits, called turbidites. They contain some chaotic, poorly sorted coarse layers at their base and then finer layers on top. The repeated sequences of turbidites indicate deposition on the continental slope and continental rise. The large landslides have occurred down the continental slope. The landslide deposits at the base of the slope are part of the continental rise. Such a landslide could trigger a tsunami.

4. Abyssal Plain

The abyssal plain is the deep ocean floor. It is basically flat, and is covered by very fine-grained sediment. It consists primarily of clay and the shells of microscopic organisms (such as foraminifera, radiolarians, and diatoms). The abyssal plain sediments may include chalk, diatomite, and shale, deposited over the basaltic ocean crust. In the deep sea, out on the abyssal plains, the depth to the seafloor varies from about 2.5 to 6 km (2500 to 6000 meters) or more below sea level. The abyssal plains receive very little sediment from the continents. The biogenic oozes: Calcareous oozes from deposits of single-cell, microscopic organisms with calcite shells result in finely laminated limestone. The siliceous oozes from single-cell, microscopic organisms with silica shells form finely laminated chert (silica) layers. Furthermore, the limestones indicate warm water; limestone dissolves in cold water. The chert indicates high biological productivity and cool water.

New Words and Phrases

1. facies *n.* 相，沉积相
2. fluvial *adj.* 河流的，河成的
3. alluvial *adj.* 洪积的
4. glacial *adj.* 冰川的
5. lacustrine *adj.* 湖泊的，湖成的
6. aeolian *adj.* 风成的
7. deltaic *adj.* 三角洲的
8. lagoonal *adj.* 潟湖的
9. estuary *n.* 河口(湾)，海湾
10. abyssal *adj.* 深海的
11. tidal flat 潮坪，潮滩
12. barrier bar 障壁坝
13. submarine canyon 海底峡谷
14. braid bar 网状河沙坝

Lesson 16

Lesson 16 ▸ Folds

Folds are defined as the undulated bends in the laminated strata of sedimentary, volcanogenic and metamorphic rocks formed in the course of plastic deformations. The series of folds comprise what is referred to as folding. The bends which have the form of folds can also appear within the process of rock formation, as is the case with the structures enveloping the irregularities of **the surfaces of unconformity**, or during the movement of lavas in the process of their solidification.

Rocks **crumpled** into folds are of the most various age and are distributed over extremely vast territories. They are exposed on the surface in Central Asia, **Kazakhstan**, in the East of **Russia**, in **the Ukraine**, **the Caucasus** and elsewhere. But even in those places where horizontally or inclined bedded strata spread, their bed or **foundation** consists of rocks crumpled into folds.

Hence, the folded forms of bedding in **the continental Earth's crust** are distributed almost everywhere.

There is every reason to say that folding is the result of plastic deformations of rocks and is developed only in **laminated strata**. In the rocks characterized by **massive composition** (e. g. **intrusive rocks**) folds are not formed, and plastic deformations find their expression in the other forms.

Plastic deformations, causing the bendings of the layers into folds, reflect the **tensions** in the Earth's crust, conditioned mainly by **endogenic processes**, the most important of which is **tectonic movement**. **The process of folding** itself can hardly be imagined without **the dislocation rocks**. This is promoted by the **lamination surfaces** that make it possible for the layers to be dislocated in the course of bending and create a regular arrangement of folds with the **alternation** of the bendings directed oppositely.

The classification of folds is based on various principles. It can rest either on the form of folds or their origin. The classification in which the folds are divided according to their form is called morphological; the classification, reflecting the conditions in which folds are formed, is referred to as genetic. The morphological and the genetic classifications consider various properties of folds and, hence, are supplementary to each other than otherwise.

The **morphological classification** presupposes the division of folds according to a number of features. According to **the position of axial surface**, the kinds of folds are **symmetrical folds**, **asymmetrical folds**, **inclined folds**, **overturned folds**, **recumbent folds**. According to the relationship between the limbs, the following types of folds will be divided by **normal folds**, **isoclinal folds** and **fan-shaped folds**. In accordance with the form of curve, folds are distinguished by **sharp folds**, **blunt folds**, **coffer (box) folds**. According to the degrees of thickness of the layers on limbs and in roofs, the following types of folds are singled out by **similar folds**, **concentric folds**, **anticlinal folds** (**supratenuous folds**) and **synclinal folds**.

According to the ratio of **the long axis of fold** (length) to its short axis (width), folds are known to be either linear, **brachy-formed** or **dome-shaped**.

Linear folds are those with which the ratio of the length and the width is greater than 3. **Folds of the oval shape** with which the said ratio is smaller than 3 are called brachy-morphic (brachy anticlines and brachy synclines). In the case when **the cross dimensions** are approximately the same, folds are called dome-shaped, while the synclinal folds of this type are referred to as **bowl-like folds** or **troughs**.

The position of the folds in the Earth's crust is greatly influenced by their **hinges**. On the Earth's surface, when **hinges of fold** are parallel, the limbs are parallel to **the axial line**. In those places where hinge of fold is either pitched or elevated, the layers **go round** the axial line; the parts of the anticlinal folds, on which the hinge of fold is inclined, are called **the periclinal closure**. The layers on these parts of the territory dip in the direction departing from **the core of the fold**. In synclinal folds the parts of the folds characterized by an inclined hinge of fold are called **centroclinal closure**. In this case the layers going round **the fold axis** are inclined towards its core.

In the rocks that have undergone **intensive regional metamorphism** and that have been transformed into **schists** and **gneisses**, it is not infrequent that we can find folds, the bends of which are situated vertically. The angles of dip of the layers on the limbs of these folds are usually either steep or equal to 90°.

The process involving the emergence and the development of folds in the Earth's crust is highly complicated and varied. At present, by far not all the aspects of this process have been made clear and it is only tentatively that we can speak of what lies behind the process of folding, adducing for this purpose the phenomena that we already know and that are connected with the deformations of rocks, or going by the facts pertaining to the historical development of the Earth's crust.

It should be pointed out that until recently a large number of those associated with the research into the problems of folding have tried to find the solution that would be universal and at the same time could furnish an unequivocal explanation of the origin of various forms of folds. Even in those scientific works where a number of factors were acknowledged to have been responsible for the formation of folds, one basic factor was recognized as the predominant one with all others considered subservient. However, the experience accumulated in the course of geological quests has provided a sufficient amount of evidence that the processes of folding are manifold and can hardly be related to one particular factor; they reflect various aspects in the development of the Earth's crust and function within a wide range of intensity, both in time and space, and are wholly dependent on the physical properties of rocks and the environment. Two **genetic classifications** are discussed below. The first classification is based on the differences in the dynamic conditions of plastic deformations; The second classification reflects the **geological medium** in which the folds are developed.

Geological conditions under which folding takes place are varied in many respects. The most widely spread folds are those that are connected with endogenic processes. In this case there emerges endogenic folding, or that of tectonic origin. **Considerably less often**, mainly in **the uppermost part of the Earth's crust**, there originate folds conditioned by **exogenic processes**, or the folding of **non-tectonic genesis**.

New Words and Phrases

1. fold *n.* 褶皱
2. crumple *vt.* 变褶，弄褶
3. foundation *n.* 基岩
4. tension *n.* 应力状态
5. alternation *n.* 互层，交替
6. trough *n.* 槽，沟
7. hinge *n.* （褶曲）枢纽
8. schist *n.* 片岩，结晶片岩
9. gneiss *n.* 片麻岩
10. genesis *n.* 成因，起源
11. Kazakhstan 哈萨克斯坦
12. the Ukraine 乌克兰
13. the Caucasus 高加索
14. the laminated strata 成层状岩层
15. the surfaces of unconformity 不整合面
16. the continental Earth's crust 大陆地壳
17. massive composition 大块岩体(岩基)
18. intrusive rocks 侵入岩
19. endogenic processes 内营力作用
20. tectonic movement 构造运动
21. the process of folding 褶皱过程
22. the dislocation rock 岩石的变形(位)
23. the lamination surface 岩层层面
24. the morphological classification 形态分类
25. the genetic classification 成因分类
26. the position of axial surface 轴面产状
27. symmetrical folds 对称褶皱
28. asymmetrical fold 不对称褶皱
29. inclined fold 倾斜褶皱
30. overturned fold 倒转褶皱
31. recumbent fold 平卧褶皱
32. normal fold 正常褶皱
33. isoclinal fold 等斜褶皱
34. fan-shaped fold 扇形褶皱
35. sharp fold 尖棱褶皱
36. blunt fold 钝状褶皱
37. coffer (box) fold 箱状褶皱
38. similar fold 相似形褶皱
39. concentric fold 同心状褶皱
40. supratenuous fold 顶薄形褶皱
41. anticlinal fold 背斜形褶皱
42. synclinal fold 向斜形褶皱
43. the long axis of fold 褶皱的长轴
44. linear fold 线状褶皱
45. brachy-formed fold (brachy-morphic fold) 短轴褶皱
46. dome-shaped fold 穹窿构造
47. fold of the oval shape 椭圆形褶皱
48. the cross dimensions 纵横维度
49. bowl-like fold 似碗状褶皱
50. hinge of fold 褶皱枢纽
51. the axial line （褶皱）轴线
52. go round 旋转，转动
53. the periclinal closure 穹隆构造闭合
54. the core of the fold 褶皱的核部

55. centroclinal structure (centroclinal closure) 同心向斜构造
56. centroclinal fold 向心褶皱
57. the fold axis 褶皱的轴
58. the geological medium 地质介质
59. intensive regional metamorphism 强烈区域变质作用
60. the uppermost part of the Earth's crust 地壳最上部
61. Considerably less often 相当少, 很少
62. non-tectonic genesis 非构造成因
63. exogenic processes 外营力作用

Lesson 17

Lesson 17 ▸ Faults

A **fault** is a fracture, or a zone of several **fractures**, across which movement has taken place, as shown by the fact that a reference marker of some kind has been offset. As such, faults differ from **joints**, which exhibit no measurable displacement. In the upper crust, above a depth of about 15 km, fault displacement is the principal means by which energy generated at **plate margins** is **dissipated**. Understanding the geometrical expression of faults, and the mechanisms by which they develop, is crucial to a number of key needs of modern society, such as the exploitation of **fault-related energy** and **mineral reserves**, and the prediction of **earthquake hazards**.

This chapter presents a description of ruptures along which considerable faults of rocks **adjoining** the surface of **rupture** took place. The first thing that should be said here is that it is impossible to draw distinct line between **fissures and ruptures**. In those places where only **fissures** are developed, a fault measuring 10 cm will be noticeable and consequently distinguished, whereas in the areas characterized by a marked distribution of faults, the rupture mentioned above will be referred to as fissures.

The classification of faults has been elaborated on the basis of the experience gained by geologists in the period of many years. These faults are divided into six main groups: **normal faults**, **reverse faults**, **shear-faults**, **gaping faults**, **overthrusts** and **nappes**. The faults of each of these groups have distinct morphological features and are formed in various dynamic and kinematic conditions. Hence, the given classification may be regarded as both morphological and genetic.

Normal faults are the deformations in which **the surface of fault** is inclined in the direction where the subsided rocks are situated.

Upthrusts are the **dislocations** in which **the surface of disruption** is inclined in the direction where the uplifted rocks are situated. The classification of upthrusts practically coincides with that of faults. Upthrusts are also distinguished by a number of features.

According to the angle of inclination of the fault fissure there are: **gently sloping upthrusts** with the angle of inclination of the fault fissure equal to 30°, **steep upthrusts**, with the angle of inclination between 30° and 80°, and **vertical upthrusts**, with which the angle of inclination of the fault fissure is within the range of 80° and 90°.

Faults and upthrusts are often developed in groups covering vast areas. Those that are widely spread are the lowered and uplifted blocks of rocks divided by faults and upthrusts and that have been termed **horsts** and **grabens**. Grabens represent linear structures formed by faults (more frequently), or upthrusts, the central parts of which are lowered and composed of younger rocks on the surface than in the uplifted marginal parts. In the structuring of grabens, faults, shear–faults and upthrust shifts can take part. The depressions in the central parts of grabens occur in a step–like manner along the lines of ruptures. There are simple and complex varieties of grabens. Simple grabens are formed by two or three faults or upthrusts; the formation of complex grabens implies the participation of a large number of ruptures. Horsts comprise linear structures formed by faults, upthrusts, faults and shear–faults and upthrust shifts, the central parts of which are elevated and composed of more ancient rocks on the surface than in the marginal parts. Horsts are either simple or complex. Simple horsts are formed by two or three faults or upthrusts; the formation of complex horsts presupposes the participation of a large number of ruptures. Horsts (as well as grabens) can be developed both subsequent to the processes of sedimentation and thus being superimposed upon the structures that have already been formed and at the same time with sedimentation itself. In the latter cases the formation of sediments is realized in the marginal parts of horsts at the expense of **disintegration** and **washout** of their elevated central areas.

As has been suggested by V. V. Beloussov, the ruptures in which the displacement of limbs takes place **at a right angle** to the surface of discontinuity are called the gaping faults. The latter in each particular case are characterized by an increased gap, or **hiatus**, between the limbs of the rupture. The amplitude of a gaping fault is measured **perpendicular to** the surface of discontinuity and can be of different value. In some cases it can be as great as tens of metres, though, as a rule, it does not exceed the limits of several metres. As compared with other types of ruptures, the gaping faults are more often filled with rocks and minerals. They are associated with both the **singular large vertical dikes**, filling their fissure faults and the linear dike belts.

Shear–faults are defined as ruptures the displacements along which take place in the horizontal direction along **the strike of the fault fissure**. They comprise limbs, fault fissure, the angle of inclination of the fault fissure, and **the amplitude of displacement**.

In relation to the strike of dislocated rocks they, as is the case with faults and upthrusts, can be of longitudinal, oblique, or diagonal, and transverse varieties. Right–hand and left–hand faults are also distinguished. In order that the character of displacement could be established, the observer should stand facing the fault fissure at the point where the layer breaks away. If on the opposite limb of the shift the layer is displaced to the right of the observer, then shear fault will be referred to as the one of the right–hand variety, if it is displaced to the left, shear fault will acquire the designation opposite to this particular direction of the movement.

The types of ruptures discussed above are characterized either by a brittle kind of discontinuity or viscous decomposition of rocks without any noticeable preliminary plastic deformations or are accompanied by hardly perceptible ones.

There are, however, widely spread ruptures that emerge and develop at the same time with the formation of folds. Ruptures of upthrust structure appearing simultaneously with folding or superimposed on folded structures are called overthrusts. Overthrusts are predominantly developed in the markedly compressed inclined or **overturned folds**. Less often they complicate

the structure of smooth gently sloping folds. In the folds formed by relatively **homogeneous rocks**, overthrusts emerge in **the hinges of folds** and are oriented parallel to **the axial plane**. In **heterogeneous masses of rock**, they can be developed in the limbs of folds along the boundaries of plastic rocks. This kind of overthrusts is widely developed, for instance, among the Cretaceous and Paleogene deposits of **Eastern Carpathians**, where they are concentrated in the normal limbs of the overturned folds, in the roof or the sole of the soft Oligocene **argillites**.

New Words and Phrases

1. fault *n.* & *vt.* 断层，产生断层
2. fracture *n.* & *vt.* 断裂，破裂
3. joint *n.* 节理，裂缝
4. dissipate *vt.* & *vi.* 消散，消失
5. adjoining *adj.* 毗连的，邻近的
6. rupture *n.* & *vt.* 断裂，破裂
7. fissure *n.* 节理，裂缝，缝隙
8. nappe *n.* 推覆体，逆掩褶曲
9. overthrust *n.* 逆掩断层，上冲断层
10. upthrust *n.* 逆断层，(上)冲断层
11. dislocation *n.* 错位，变位
12. horst *n.* 地垒
13. graben *n.* 地堑
14. disintegration *n.* 崩解，解体
15. washout *n.* 冲蚀
16. hiatus *n.* 裂隙，间断
17. argillite *n.* 泥岩，泥板岩
18. plate margin 板块边缘
19. mineral reserves 矿产资源
20. earthquake hazard 地震灾害
21. fault-related energy 断层相关的能源
22. fissures and ruptures 节理与断层，裂缝与破裂
23. normal fault 正断层
24. reverse fault 逆断层
25. shear fault 剪切断层
26. gaping fault 张开断层
27. overthrusts and nappes 逆掩断层和(冲断)推覆体
28. structural trap 构造圈闭
29. the surface of fault 断层面
30. the surface of disruption 断裂面
31. gently sloping upthrust 平缓逆断层
32. steep upthrust 陡峭逆断层
33. vertical upthrust 直立逆断层
34. at a right angle 成直角
35. be perpendicular to 垂直于
36. the singular large vertical dikes 奇异硕大的垂直状岩脉
37. the strike of the fault fissure 断层面的走向
38. the amplitude of displacement 断距，位移幅度
39. overturned folds 倒转褶皱
40. the structure of smooth gently sloping folds 平缓光滑褶皱构造
41. homogeneous rock 均质岩石
42. the hinges of folds 褶皱的枢纽
43. the axial plane (褶曲)轴面
44. heterogeneous masses of rock 非均质岩体
45. Eastern Carpathians (中欧)东喀尔巴阡山脉

Lesson 18 ▸ Unconformities

The relationship between the rocks entering into **the composition of layered strata** can be exemplified by two **particular instances**. In the first case, each of the **superimposed layers** or a complex of layers, making up the given **stratigraphic horizon**, is found to be on the top of **the underlying rocks** without the slightest **indication of discontinuity** in the process of sedimentation. This kind of relationship reflecting the sedimentation process **function as a prerequisite for** the **concordant** bedding of rocks. In the second case, between the superimposed and the underlying layers, the **stratigraphic succession** is disturbed and the deposits of certain stratigraphic horizons in the **section** are absent. In this case there emerges a discordant bedding of rocks.

The origin of **unconformities** can be conditioned by various factors. They can be the result of either **a disruption in sedimentation** or they can appear during **the tectonic dislocations** of some of the strata in respect to others. In the former case the unconformities are called **stratigraphic unconformity**, in the latter case they are referred to as **tectonic unconformity**.

In stratigraphic unconformities the fallout of any of the rocks from the sections is caused by **the cessation of sedimentation**, or the change of the **regime** that proves to be favourable for the accumulation of sediments, and the conditions under which the sediments are no longer accumulated or there occurs a **disintegration** and washout of the previously formed rocks. According to a number of various features, stratigraphic unconformities can be divided into several types. The classification is based on the differences in the value of **the angle of unconformity**, the distinctness with which the surface of unconformity is expressed, the areas of distribution and the factors conditioning their emergence. A particular group comprises unconformities that are brought about in **subaqueous conditions**.

In accordance with the angle of unconformity, the latter is distinguished as either **parallel unconformity**, **angular unconformity**, or **geographical unconformity**.

Parallel unconformity is expressed by a stratigraphic interruption between the series of layers bedded above and below **the surface of unconformity** parallel to each other. The two series are usually differentiated by **the composition of rocks** , the fossils enclosed in them and other features. The surface of unconformity **delimiting** the series finds its expression in the most overt form. It usually represents **the surface of subaqueous washout** or **terrestrial denudation** formed at **the time of elevation** and when the process involving **the formation of sedimentary strata** ceased to function. Thus, for instance, during the **Carboniferous period the territory of the nearby districts of Moscow** was the accumulation zone of **marine carbonaceous silts**. At the end of the Carboniferous period, this area was elevated above sea level, and in the Permian, Triassic, and the greater part of Jurassic period it was remarkable for **the continental conditions** under which the disintegration of earlier formed rocks was taking place. The end of the **Jurassic period** was marked by the beginning of **an overall subsidence of land** which was characterized by **an expansive advance of the sea** resulting in the accumulation of the stratum containing **argillaceous silts**. These movements left their marks in the section which took the form of **stratigraphic breaks** (the absence of rocks of the Permian and the Triassic age) and a clearly expressed surface of unconformity at the border between the **Carboniferous limestones** and the Upper Jurassic clays.

Angular unconformity finds its expression in the interruption between the two complexes of layers that have a different angle of inclination. This kind of unconformity is usually clearly reflected both in the natural vertical sections and in the exposures upon the Earth's surface. Besides, the main features of angular unconformity have the same expression both in the vertical sections and on the geological maps. The surface of unconformity, dividing the unconformable **suites** discordantly in both cases, cuts off different horizons of **the ancient suite** at an angle and passes to a certain extent parallel to the horizons between the individual horizons of **the young suite**. This feature proves to be one of the most important ones in establishing angular unconformity in the process of geological mapping and when the geological maps are to be interpreted.

Presumably, stratigraphic unconformities, originating in subaqueous conditions, are developed on a much wider scale than it has so far been stated.

The discordant contacts between the layers of different ages and lithological composition can be caused by the tectonic ruptures and the dislocations along them of individual blocks of rock. In those cases when the exposure is unhampered and the geological investigations presuppose a fairly detailed analysis, it is usually not difficult to discover the stratigraphic and the tectonic unconformities and draw a clear line between them.

However, there might be complicated cases when it is difficult to correctly assess and give **a hard and fast** interpretation of geological observations. Thus, for instance, in the case of a complex folded structure the nature of the contact does not often lend itself to interpretation when it is situated parallel in respect to the general form of the bedding of **the overlying suite**.

The thorough investigation of the contact can furnish the following additional data: (1) the presence of a system of **gliding planes** and **ground masses** of **tectonic breccia** shows that the formation of the contact could largely be conditioned by tectonic factors; (2) the presence of **the basal conglomerate** in the foundation of **the upper suite** as well as the clearly expressed traces of weathering in the surface zone of **the lower suite irrefutably proves** the existence of the surface of angular unconformity separating the two suites.

There is substantial material that can present a detailed analysis of the relationship between the upper suite and the surface of contact. When the distribution of lamination and contact is parallel, the solution of the problem rests upon the recognition of **the transgressive bedding** of the upper suite. When **the contact plane** cuts off the layers of the upper suite, the presence of

tectonic contact is found to be more feasible. However, in the present case **there can still be room for doubt**, since this kind of relationship between **the lamination of the suite** and its sole can emerge during the disharmonic, intrabedded deformations, that occur particularly often when there is a marked difference in the lithological composition of two suites of different ages, as well as in the case of **contiguity**.

New Words and Phrases

1. unconformity *n.* 不整合
2. superimpose *vt.* 使重叠，使叠加
3. concordant *adj.* 整合的，一致的
4. section *n.* （地层）剖面
5. cessation *n.* 中断，间断
6. regime *n.* 沉积环境
7. disintegration *n.* 崩解，解体
8. delimit *vt.* 划定，界定
9. suite *n.* 岩系
10. contiguity *n.* 接近，邻近
11. the composition of layered strata 层状岩层
12. particular instance 特例
13. stratigraphic horizon 地层层位
14. superimposed layers 叠加岩层
15. the underlying rock 下伏地层
16. indication of discontinuity 无（地层）缺失显示
17. function as a prerequisite for 作为一个先决条件
18. stratigraphic succession 地层层序
19. a disruption in sedimentation 沉积间断
20. the tectonic dislocation 构造变位
21. stratigraphic unconformity 地层不整合
22. tectonic unconformity 构造不整合
23. the cessation of sedimentation 沉积间断
24. the angle of unconformity 不整合角度
25. subaqueous condition 水下条件
26. parallel unconformity 平行不整合
27. angular unconformity 角度不整合
28. geographical unconformity 地理不整合
29. the surface of unconformity 不整合面
30. the composition of rock 岩石组合
31. the surface of subaqueous washout 水下冲刷面
32. the surface of terrestrial denudation 地表剥蚀面
33. the time of elevation （地壳）抬升期
34. the formation of sedimentary strata 沉积岩层的形成
35. the territory of the nearby districts of Moscow 莫斯科附近区域
36. Carboniferous period 石炭纪
37. marine carbonaceous silt 海相碳质泥沙
38. the continental condition 大陆条件（环境）
39. Jurassic period 侏罗纪
40. an overall subsidence of land 陆地整体沉降
41. an expansive advance of the sea 海进
42. argillaceous silt 泥质粉砂（岩）
43. stratigraphic break 地层间断
44. Carboniferous limestone 石炭纪石灰岩
45. the ancient suite 古老岩层
46. the young suite 年轻地层
47. a hard and fast 一板一眼的，严厉的
48. the overlying suite 上覆岩层
49. gliding plane 滑动面

50. ground masses　地面岩体
51. tectonic breccia　构造角砾岩
52. the basal conglomerate　底砾岩
53. the upper suite　上部岩层
54. the lower suite　下部岩层
55. irrefutably prove　无可辩驳地证明
56. the transgressive bedding　海侵地层
57. the contact plane　(不整合)接触面
58. tectonic contact　构造面(界限)
59. there can still be room for doubt　仍然有值得怀疑的地方
60. the lamination of the suite　(一套)岩层

Lesson 19 ▸ Plate Tectonics

The theory of plate tectonics has done for geology what Charles Darwin's theory of evolution did for biology. It provides geology with a comprehensive theory that explains "how the Earth works". The theory was formulated in the 1960s and 1970s as new information was obtained about the nature of the ocean floor, Earth's ancient magnetism, the distribution of volcanoes and earthquakes, the flow of heat from Earth's interior, and the worldwide distribution of plant and animal fossils.

The theory states that Earth's outermost layer, the **lithosphere**, is broken into 7 large, **rigid pieces** called plates: the African, North American, South American, Eurasian, Australian, Antarctic, and Pacific plates. Several minor plates also exist, including the Arabian, Nazca, and Philippines plates.

The plates are all moving in different directions and at different speeds (from 2cm to 10cm per year, about the speed at which your fingernails grow) in relationship to each other. The plates are moving around like cars in **a demolition derby**, which means they sometimes crash together, pull apart, or **sideswipe** each other. The place where the two plates meet is called **a plate boundary**. Boundaries have different names depending on how the two plates are moving in relationship to each other: (1) **crashing-convergent boundaries**; (2) **pulling apart-divergent boundaries**; (3) **sideswiping-transform boundaries**.

Places where plates crash or crunch together are called convergent boundaries. Plates only move a few centimeters each year, so collisions are very slow and last millions of years. Even though **plate collisions** take a long time, lots of interesting things happen. For example, in the drawing above, an **oceanic plate** has crashed into a **continental plate**. Looking at this drawing of two plates colliding is like looking at a single frame in a slow-motion movie of two cars crashing into each other. Just as the front ends of cars fold and bend in a collision, so do the "front ends" of colliding plates. The edge of the continental plate in the drawing has folded into a huge mountain range, while the edge of the oceanic plate has bent downward and dug deep into the Earth. A **trench** has formed at the bend. All that folding and bending makes rock in both plates break and slip, causing earthquakes. As the edge of the oceanic plate digs into **Earth's hot interior**, some of the rock in it melts. The melted rock rises up through the continental plate, causing more earthquakes on its way up, and forming **volcanic eruptions** where it finally reaches the surface. An example of this type of collision is found on the west coast of South America where **the oceanic Nazca Plate** is crashing into the continent of South America. The crash formed the **Andes Mountains**, the long string of volcanoes along **the mountain crest**, and the deep trench off the coast in the Pacific Ocean.

Mountains, earthquakes, and volcanoes form where plates collide. Millions of people live in and visit the beautiful mountain ranges being built by plate collisions. For example, the Rockies in North America, the Alps in Europe, **the Pontic Mountains in Turkey**, **the Zagros Mountains** in Iran, and the Himalayas in central Asia were formed by plate collisions. Each year, thousands of people are killed by earthquakes and volcanic eruptions in those mountains. Occasionally, big eruptions or earthquakes kill large numbers of people.

Places where plates are coming apart are called divergent boundaries. As shown in the drawing above, when Earth's brittle surface layer (the lithosphere) is pulled apart, it typically breaks along parallel faults that tilt slightly outward from each other. As the plates separate along the boundary, the block between the faults cracks and drops down into the soft, plastic interior (the asthenosphere). The sinking of the block forms a central valley called a **rift**. Magma seeps upward to fill the cracks. In this way, new crust is formed along the boundary. Earthquakes occur along the faults, and volcanoes form where the magma reaches the surface.

Plate separation is a slow process. For example, divergence along **the Mid Atlantic ridge** causes the Atlantic Ocean to widen at only about 2 centimeters per year.

Places where plates slide past each other are called transform boundaries. Since the plates on either side of a transform boundary are merely sliding past each other and not tearing or crunching each other, transform boundaries lack the spectacular features found at convergent and divergent boundaries. Instead, transform boundaries are marked in some places by **linear valleys** along the boundary where rock has been ground up by the sliding. In other places, transform boundaries are marked by features like **stream beds** that have been split in half and the two halves have moved in opposite directions.

Although transform boundaries are not marked by spectacular surface features, their sliding motion causes lots of earthquakes. The strongest and most famous earthquake along **the San Andreas fault** hit San Francisco in 1906. Many buildings were shaken to pieces by the quake, and much of the rest of the city was destroyed by the fires that followed. More than 600 people died as a result of the quake and fires. Recent large quakes along the San Andreas include the Imperial Valley quake in 1940 and the Loma Prieta quake in 1989.

The main features of **plate tectonics** are: (1) the Earth's surface is covered by a series of crustal plates; (2) the ocean floors are continually moving, spreading from the center, sinking at the edges, and being regenerated; (3) convection currents beneath the plates move the crustal plates in different directions; (4) the source of heat driving the **convection currents** is radioactivity deep in the Earth's mantle.

The distribution of **mid-oceanic ridges**, **geomagnetic anomalies**, **deep sea trenches** and **island arcs**, all these observations, along with many other studies of our planet, support the theory that underneath the Earth's crust is **a malleable layer** of heated rock known as the **asthenosphere** which is heated by radioactive decay of elements such as Uranium, Thorium, and Potassium. Because the radioactive source of heat is deep within the mantle, the fluid asthenosphere circulates as convection currents underneath the solid lithosphere. This heated layer is the source of lava we see in volcanos, the source of heat that drives hot springs and geysers, and the source of raw material which pushes up the mid-oceanic ridges and forms new ocean floor. Magma continuously wells upwards at the mid-oceanic ridges (arrows) producing currents of magma flowing in opposite directions and thus generating the forces that pull the sea floor apart at the mid-oceanic ridges. As the ocean floor is spread apart, cracks appear in the middle of the ridges allowing molten magma to surface through the cracks to form the newest ocean floor. As the ocean floor moves away from the mid-oceanic ridge it will eventually come into contact with a continental plate and will be subducted underneath the continent. Finally, the lithosphere will be driven back into the asthenosphere where it returns to a heated state.

New Words and Phrases

1. lithosphere *n.* 岩石圈
2. sideswipe *vt.* 擦边撞击，沿边擦过
3. trench *n.* 海沟，深沟
4. rift *n.* 裂缝，裂隙
5. asthenosphere *n.* 软流圈，岩流圈
6. plate tectonics 板块学说
7. rigid pieces 刚性板块
8. Nazca 纳斯卡
9. a demolition derby 一场破坏性比赛
10. a plate boundary 板块边缘(界)
11. crashing-convergent boundaries 碰撞—聚合型边界
12. pulling apart-divergent boundaries 拉张—分离型边界
13. sideswiping-transform boundaries 剪切—转换型边界
14. plate collision 板块碰撞
15. oceanic plate 大洋板块
16. continental plate 大陆板块
17. Earth's hot interior 热的地球内部
18. volcanic eruption 火山喷发
19. the oceanic Nazca Plate 纳斯卡大洋板块
20. Andes Mountains 安第斯山脉
21. the mountain crest 山脊
22. the Pontic Mountains in Turkey 土耳其桥体山
23. the Zagros Mountains 扎格罗斯山脉(伊朗)
24. plate separation 板块分裂(离)
25. the Mid Atlantic ridge 中大西洋海岭
26. linear valleys 线状山谷
27. stream bed 河床
28. the San Andreas fault 圣安德烈亚斯断层
29. convection currents 对流
30. mid-oceanic ridges 大洋中脊
31. geomagnetic anomalies 地磁异常
32. deep sea trenches 深部海沟
33. island arcs 岛弧
34. a malleable layer 可塑层

Lesson 20 ▸ Tectonic Movements

Tectonic movements are understood as the dislocations of a substance that reflects the development of the Earth's crust and **the abyssal mantles** of our planet. They are divided into **contemporaneous tectonic movements**, the **recent tectonic movements**, and those that refer to the earliest period, or the **ancient tectonic movements**. The contemporaneous movements are determined by the readings of the instruments, historical data, **archaeology**, or by the registered changes in the relief. The recent movements are most closely related to the formation of the contemporaneous relief. Their lower age limit does not have any particular date mark, though it coincides with the beginning of Oligocene or Neogene period.

All the three types of movements, the contemporaneous, recent and ancient movements comprise an uninterrupted chain reflecting, as it were, events on different scales and, consequently, with a different degree of detailedness. The latter, which we find accessible to attain in studying the contemporaneous movements, becomes blurred in our analysis of the recent movements and is completely beyond our perception when it comes to the reconstruction of the ancient movements.

The contemporaneous and recent movements are characterized by the velocities and trajectories of the displacement of some points or lines that are chosen on the surface of the Earth or at the depth. In the reconstruction of ancient movements we make extensive use of the principle of actualism, the facies analysis, the study of the degrees of thickness, etc.

Data concerning the vertical constituents of tectonic movements are absolutely reliable. The average velocities of the recent movements in **orogenic rocks** made 0. 3 mm per year, while on the platforms they are 0. 05 mm per year. Since within one and the same period of time the resultant amplitudes of all the three types of movements are equal, while the length of the rated trajectories is different, the velocities of the recent movements are found to be smaller than those of the contemporaneous movements, and the velocities of the ancient movements are smaller than those of the recent ones.

Calculations of the average velocities of movements for individual periods of time in the Late Proterozoic, Early and Late Paleozoic, Mesozoic and Cenozoic, with the conditions given above of their determination, as well as geological data taken into account, show a systematic growth of absolute amplitudes from the ancient stages in the development of the Earth's crust to the younger ones, which makes it possible for us to form a general idea concerning the growth of velocities and amplitudes of **vertical movements** within the framework of geological time.

Investigations pertaining to the study of contemporaneous **horizontal movements** have been initiated only recently. Besides the repeated **geodetic measurements**, the first results have been provided by **special geodynamic polygons created for this purpose** in mountainous regions. In determining the global distances, no minor significance can be attributed to the artificial Earth satellites, as well as to photo–zenith and laser installations.

Tectonic movements should not be regarded entirely in their concrete expression, but also in how far they influence the formation of the Earth's crust and the conditions of excitation.

The movements are divided into essentially horizontal and those that are likewise vertical. According to the sources of excitation and the spheres of their manifestation, they are further classified into the global, upper mantle, crustal and surface varieties.

The global movements cover the whole of the Earth from the core to the surface. The upper mantle movements are originated below the sole of the Earth's crust. They are propagated over considerable segments of the Earth's crust, and are commensurable in area with **the geosynclinal regions**. The crustal movements are manifested in the Earth's crust and are particularly intensive above **Conrad discontinuity**. As far as its properties are concerned **the basaltic layer** that is situated below has much more in common with **the supraasthenospheric layer** than with **the overlying granitic layer**. The crustal movements are conditioned by **the displacements of magma**, disturbances of isostatic equilibrium, instability of the Earth's magnetic field, displacements and turnings of the large parts of continents. The crustal movements are genetically connected with the linear, folded and ruptural deformations.

In assessing the function of different types of tectonic movements in the formation of the Earth's crust we should not overestimate the role of either vertical or horizontal movements. We could hardly be justified in regarding one of these types of movements as primary and the other as secondary, or the one that is subservient to its counterpart. Both these types, the vertical and the horizontal movements have their own origin though they are closely interconnected, and in certain forms can also be derived from one and the same source.

Vertical movements may well be regarded as factors conditioning the distribution of regions of **denudation and sedimentation** on the surface of the Earth. These movements accompany **magmatic processes**, **gravitational phenomena**, and the formation of various forms of folds and ruptures.

Horizontal movements are brought about by a large number of factors. No minor significance can be attributed to the disturbances in **the rotation regime of the planet**. There are substantiated **assumptions** concerning different velocities of the rotation of the core and the mantles of the Earth. Horizontal movements are associated with the formation of folded and ruptural structures in the geosynclinal regions and, possibly, with the formation of magmas as well.

With all the variety of factors responsible for the movements of the Earth's crust, it would be remiss on our part not to mention the fact that the basic energy excitation levels of vertical and horizontal movements are possible beyond the boundaries of the lithosphere, in the upper mantle. At the same time it would hardly be feasible not to regard the changes that continually occur within all the three layers of the Earth's crust. The general direction in which the Earth's crust develops is presumably determined mainly by the processes that occur in the mantle, which find their reflection in the Earth's crust, and to which the Earth's crust is continually accommodating itself through constructive and destructive processes.

New Words and Phrases

1. archaeology *n.* 考古学
2. assumption *n.* 假定，假设
3. tectonic movement 构造运动
4. the abyssal mantle 深部地幔
5. contemporaneous tectonic movement 现代构造运动
6. recent tectonic movement 近代构造运动
7. ancient tectonic movement 古代构造运动
8. orogenic rock 造山带岩石
9. vertical movement 垂直运动
10. horizontal movement 水平运动
11. geodetic measurement 大地测量
12. special geodynamic polygon 特殊地球动力学多边形
13. created for this purpose 为此目的创建
14. the geosynclinal region 地槽区
15. Conrad discontinuity 康拉德界面
16. the basaltic layer 玄武岩层
17. the supraasthenospheric layer 上软流圈层
18. the overlying granitic layer 上覆花岗岩质层
19. the displacements of magma 岩浆的移动
20. magmatic processes 岩浆作用
21. denudation and sedimentation 剥蚀作用与沉积作用
22. gravitational phenomena 重力现象
23. gravitational slide phenomena 重力滑动现象
24. the rotation regime of the planet 地球自转机制

Lesson 21 ▸ Styles of Basin Structures

Contractional Folds and Thrust-fault Structures

Contractional folds occur in areas undergoing tectonic shortening, and are generally associated with convergent plate boundaries, particularly where continent-continent collision has taken place. They may also develop where transpression occurs along **strike - slip boundaries**. A thrust fault is a type of fault, or breaks in the Earth's crust across which there has been relative movement, in which rocks of lower stratigraphic position are pushed up and over higher strata. They are often recognized because they place older rocks above younger.

Anticlines may also develop in areas of local contraction along strike-slip systems. The Wilmington Field in the Los Angeles Basin, California, is such an example. It is developed along the San Andreas Fault system. Transpressional anticlines tend to be arranged **en echelon** and are very strongly faulted, they depend on the presence of thick caprocks to seal reservoirs across fault planes.

Some contractional anticlines have developed as a result of the reversal of movement along old extensional faults. These are known as inversion anticlines. The evidence for the earlier extensional history is usually the thickening of sediments towards the fault plane during its period of growth. Good examples are the so-called Sunda folds of some Indonesian basins, for example, central Sumatra. Inversions tend to occur in areas where relatively subtle changes in the regional stress field cause reversal of movement along faults, and are therefore frequently associated with fault systems that have a strong strike-slip component.

Extensional Structures

Extensional structures form a very important group of traps, being responsible for many of the fields discovered in basins that have experienced a phase of rifting in their geological history. We will deal in this section only with structural traps resulting from extension of the basement, that is, in **stretched rift basins**. The extensional structures occurring, for example, in delta sequences that developed in the post-rift stage on **passive margins** is covered in next headline on gravitational structures.

Rollover anticlines may develop in association with basement-controlled growth faults. The Vicksburg flexure in south Texas is an example. Very large sediment thickness changes occur across the fault zone, particularly in the Oligocene section. Large quantities of oil and gas are trapped in rollover anticlines, fault traps, and stratigraphic pinch-outs.

Gravitational Structures

An important **gravitational structure** that forms petroleum traps is the rollover anticline into listric **growth faults**, occurring particularly in delta sequences. These structures are not caused by extension in the basement, but are due to instability in the sedimentary cover and its movement under gravity. They are most prone to form where a level of undercompacted (overpressured) clays or **ductile salt** occurs at depth, into which the growth faults sole out, and which is overlain by a thick sequence of more competent rocks. These conditions are commonly created in thick, **progradational delta sequences**, as in the Gulf of Mexico and Niger Delta.

The **dip-closed roll-over anticline** (Figure 21.1) is the least risky trap for petroleum, but growth faulting may also give rise to fault-closed traps. The integrity of the fault trap depends on the juxtaposition of a shale seal across the fault plane, so these traps tend to be effective on those parts of the delta where the sand/shale ratio is relatively low (say, less than 50%). Even here, many of the sands may be water-bearing owing to cross-fault leakage; production may be obtained from relatively few sands, interspersed with the water-bearing zones, and distributed over a large gross vertical interval.

Compactional Structures

The most important trap type formed by **compactional processes** is the drape anticline, caused by differential compaction. Basement (effectively non-compactible) topography causes significant thickness variations in the overlying highly compactible sediments, which compact and subside most where they are thickest off-structure. If the area above the horst remains elevated relative to surrounding areas, shallower-water sedimentary facies may develop which are less compactible than the surrounding muds. This will exaggerate the differential compaction. The sedimentary facies and diagenetic history of the reservoir unit may be more different over the crest of the drape anticline than off its flanks.

Drape anticlines form a very successful trap type. They are frequently simple features formed without tectonic disturbance (unless basement faults are reactivated) and frequently persist over a long period of geological time, from shortly after the time of reservoir deposition through to the present day.

Examples of drape anticlines are the Forties and Montrose fields of the North Sea, which occur within very large (90km^2 and 181km^2 respectively) low-relief domal closures at Paleocene level formed by drape over deeper fault blocks. Drape anticlines may indeed be very large (the world's largest field, Ghawar in Saudi Arabia, comprises Jurassic carbonates draped over a basement high), while the drape structures over the small Devonian pinnacle reefs of the Western Canada sedimentary basin are examples at the other extreme.

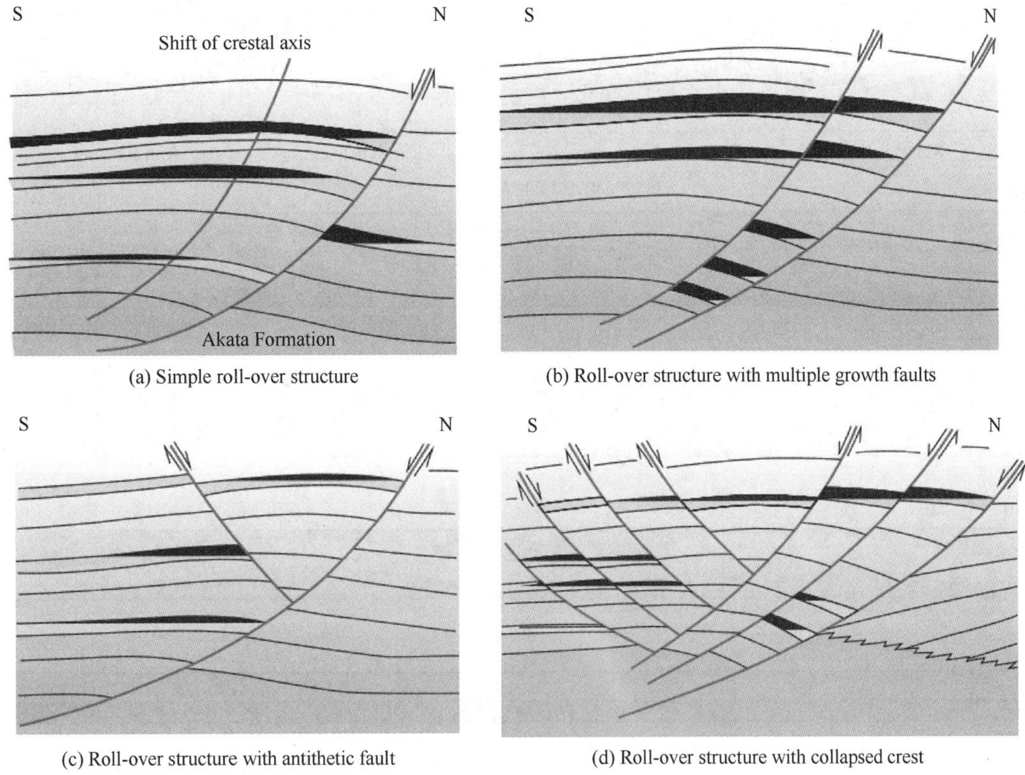

(a) Simple roll-over structure

(b) Roll-over structure with multiple growth faults

(c) Roll-over structure with antithetic fault

(d) Roll-over structure with collapsed crest

Figure 21. 1 Varieties of roll-over structure forming petroleum traps in the Niger Delta

Diapiric structures result from the movement of salt or overpressured clay. At depths in excess of 600~1000m, salt is less dense (2200kg/m^3) than its overburden (typically 2500~2700kg/m^3), is rheologically exceptionally weak, and consequently liable to lateral and upward movement. Salt can flow at surprisingly low temperatures and over long periods of geological time. Once a density inversion is present, heterogeneities in either the mother layer of salt or clay, or in the overburden, are sufficient to trigger upward movement. Examples of heterogeneities are lateral changes in thickness, density, viscosity or temperature. These changes may be essentially depositional or may be imposed as a result of faulting or folding. In extensional faulted zones, diapirs tend to form through buoyancy where overburden load is most reduced in the footwall. The salt rollers of the US Gulf Coast are examples of this triggering mechanism. Faulting may be basement-involved, or thin-skinned, usually soling out in the ductile layer. This layer may also provide a zone of detachment in contractional areas. Differential loading of a salt layer by thick overlying sediments is a powerful triggering mechanism of diapirism in young shallow delta sequences.

In simple terms, salt structures pass through three stages of growth (Figure 21. 2) —pillow stage, diapir stage and post-diapir stage.

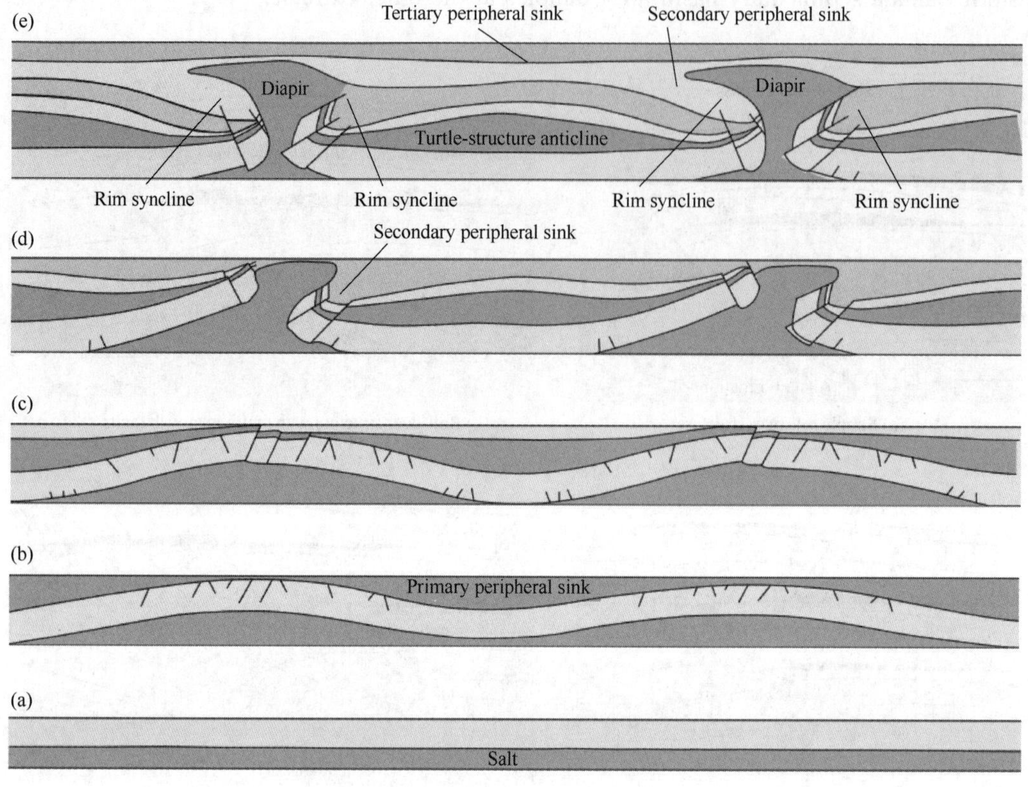

Figure 21.2 Evolution of salt structures through pillow stage (b),(c), diapir stage (d), and post-diapir stage (e)

New Words and Phrases

1. contractional fold 挤压褶皱，收缩褶皱
2. thrust-fault structures 冲断构造
3. strike-slip boundaries 走滑边缘(区)
4. strike-slip fault systems 走滑断层系
5. en echelon 雁列状(式)
6. extensional structures 伸展构造
7. stretched rift basins 伸展裂谷盆地
8. passive margins 被动(大陆)边缘
9. gravitational structure 重力构造
10. growth faults 生长断层
11. ductile salt 韧性盐
12. progradational delta sequences 进积三角洲层序
13. dip-closed roll-over anticline 倾闭式滚动(翻转)背斜
14. compactional structures 压实构造
15. drape anticline 披覆(式)背斜
16. diapiric structures 底辟构造

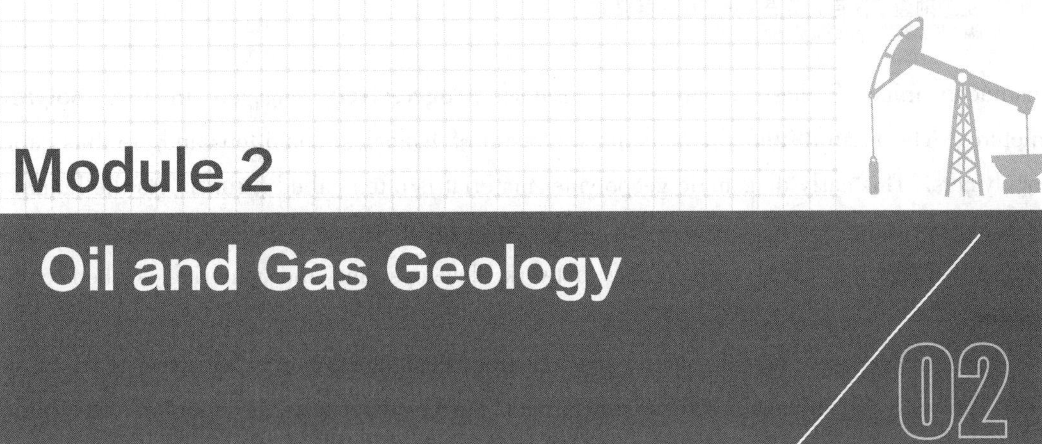

Module 2

Oil and Gas Geology

Lesson 22 ▸ Origin of Oil and Gas

Once an **organism** dies, the process of petroleum generation begins. The majority of organic matter is totally oxygenated and degrades to carbon **dioxide**, water and small amounts of **mineral** matter. The organic matter that is deposited in a "reducing" environment suffers only minor **oxidation**. The process of petroleum generation is divided into three stages—diagenesis, **catagenesis** and **metagenesis**. These divisions are artificial, since the process is a continuous one, without clear-cut boundaries. However, a number of different events do occur in each of the three stages, thus permitting a useful **demarcation** of stages. The process is one where biologically produced molecules tend to move to lower energy levels, and a final state of **equilibrium**.

Diagenesis

The first stage in the transformation of freshly deposited organic matter into petroleum is called **diagenesis**. This process begins at the sedimentary interface and extends to varying depths, but usually no deeper than a few hundred meters. In cases where the **geothermal gradient** is extremely low, diagenesis may extend to a depth of 2,000 meters. During this time compaction of the sediments occurs, the temperatures and pressures are low, causing transformation to begin under mild conditions.

During early diagenesis, one of the main agents of transformation is **microbial** activity. Depending on the oxygen content of the sea water and sediments, microbial transformation of organic matter is either **aerobic** or **anaerobic**. Biological polymers (lipids, proteins, etc.) are destroyed by microbial activity and mild chemical reactions occur during this time. The constituent units of these biopolymers become **progressively** engaged in new **polymer** structures. The **recombined** polymers are the result of geological conditions and are thus called geopolymers. This early diagenetic geopolymer material is often called **humin**. As the humin is buried deeper by increasing overburden, it becomes progressively more polymerized and more chemically **inert**. A large carbon ring network develops and the material is then called **kerogen**.

During diagenesis, organic matter loses a great deal of oxygen in the form of H_2O and CO_2. The only significant hydrocarbon formed during diagenesis is microbial **methane**. Diagenesis causes a decreasing O/C, with only a slight decrease in H/C. The end of diagenesis is generally considered to be when the vitrinite component of kerogen exhibits an optical reflectance of 0.5%, under oil **immersion**. This also corresponds to a Rock Eval T_{max} value of approximately 410℃ to 420℃.

Catagenesis

As burial continues, the kerogen formed during diagenesis is exposed to increasing temperatures and pressures. Catagenesis is the stage of thermal degradation of kerogen that forms oil and gas. This stage typically occurs between the depth of several hundred to several thousand meters.

Sediments are further compacted, expelling water and altering the **clay mineral** fraction. The **organic fraction** undergoes major alteration during catagenesis. As a result of the temperature, kerogen is "cracked" to form liquid petroleum and gas. This **cracking** of kerogen is the breaking of the carbon-carbon bonds in the **polymeric** network. It is mainly the aliphatic side-chains which are cleaved off from the main aromatic kerogen network. It is these side-chains which form petroleum. Later stages of catagenesis results in the formation of methane from kerogen and also some cracking of already formed petroleum.

The main chemical change experienced during catagenesis is a decrease in H/C, resulting from the formation of hydrocarbons. The hydrogen portions of kerogen are released during the formation of petroleum, leaving behind a residue rich in carbon. This is the reason that oil-prone kerogens are rich in hydrogen relative to carbon and oxygen.

The end of catagenesis is generally accepted to be when all the major side chains of kerogen have been cracked, and the remaining carbon network begins to take on **mineralogical** ordering of parallel **aromatic** sheets. This generally occurs at vitrinite reflectance levels of approximately 2% and T_{max} values between 480℃ and 490℃. It is within catagenesis that the majority of petroleum is formed. This is the "oil window".

Metagenesis

The metagenesis stage is reached at great depths, or in areas of high geothermal gradients at shallower depths. Metagenesis usually begins at depths of approximately 4,000 meters. At this stage, kerogen has very little hydrogen remaining and is forming methane as its only hydrocarbon product. Towards the end of metagenesis, virtually no hydrocarbons are being generated from the kerogen. Throughout metagenesis, the **residual carbon** network takes on an increasingly ordered structure, where aromatic kerogen rings are condensed into parallel plates, as in **graphite**.

The H/C ratio and hydrogen index decrease only slightly during metagenesis, since most of the hydrocarbons have already been generated. The completion of metagenesis occurs at vitrinite reflectance values around 4% and T_{max} values above 510℃.

The Oil Window

The formation of oil from kerogen depends on the amount and type of source material present in the sediments and the thermal history of the kerogen. It has been conclusively determined that temperature is the most important factor affecting the generation of oil and gas from organic matter.

During the generation of petroleum from kerogen, both temperature and time play key roles. Kerogen, exposed to relatively high temperatures for a short period of time, will mature to about the same extent as kerogen which is exposed to relatively low temperatures for a longer period of time. Thus, the time and temperature history of a kerogen determines occurrence and depth at which kerogen generates oil or gas or both.

The depth range over which oil generation occurs is known as the "oil window". This oil window is usually different for most sedimentary basins. It may cover several thousand meters or may be confined to less than a thousand meters. The depth range for the oil window is mainly a function of the past geothermal gradient and is similar to the present geothermal gradient if the subsidence history is continuous. If this assumption is correct, then the oil window can be estimated from the current downhole temperatures. However, if the tectonic history of the basin has been active, then such assumptions can lead to gross errors. The determination of the oil window is best performed by geochemical means using T_{max}, bitumen extraction, gas chromatography and optical methods such as vitrinite reflectance.

In the search for petroleum reservoirs, accurate determination of the oil window is important, because if an organically rich, oil-prone, source rock has not reached the oil window range of temperatures, it will not generate oil. Determination of the geologic time at which oil was generated within a reservoir is an important factor in evaluating the possibility of the presence of suitable structures to accumulate and trap the oil. If the generation of oil occurred prior to the formation of suitable reservoirs and traps, the likelihood of finding commercial quantities of oil is less certain, than if the reservoir and trap existed prior to the generation and migration of the oil.

New Words and Phrases

1. organism *n.* 生物体,有机体
2. dioxide *n.* 二氧化物
3. mineral *n.* 矿物
4. oxidation *n.* [化]氧化
5. catagenesis *n.* 深成热解作用
6. metagenesis *n.* 沉积变质作用
7. demarcation *n.* 划分
8. equilibrium *n.* 平衡
9. diagenesis *n.* [地质]成岩作用
10. geothermal *adj.* 地热的,地温的
11. gradient *n.* 梯度
12. microbial *adj.* 微生物的
13. aerobic *adj.* 依靠氧气的,与需氧菌有关的
14. anaerobic *adj.* [微]没有空气而能生活的,厌氧性的
15. progressively *adv.* 日益增多地
16. polymer *n.* 聚合体
17. recombined *adj.* 调配(制)的
18. humin *n.* 胡敏素
19. inert *adj.* 惰性的
20. kerogen *n.* 干酪根
21. methane *n.* [化]甲烷,沼气
22. immersion *n.* 沉浸
23. clay mineral 黏土矿物
24. organic fraction 有机馏分
25. cracking *n.* 裂解
26. polymeric *adj.* 聚合的,聚合体的
27. mineralogical *adj.* 矿物学的
28. aromatic *adj.* 芳香的
29. residual carbon 残余碳
30. graphite *n.* 石墨

Lesson 23

Lesson 23 ▸ Source Rock

A petroleum source rock is defined as any rock that has the capability to generate and expel enough hydrocarbons to form an accumulation of oil or gas. A potential source rock is one that is too **immature** to generate petroleum in its natural setting but will form significant quantities of petroleum when heated in the laboratory or during deep burial. An effective source rock is one that has already formed and expelled petroleum to a reservoir. It may be active (currently expelling) or inactive (e. g., because of uplift with **erosion** and cooling). Also, crude oil-source rock correlations have helped establish that a particular rock has yielded oil to a particular reservoir.

The term kerogen originally referred to the organic matter in oil **shales** that yielded oil upon heating. Subsequently, the term was defined as all the disseminated organic matter of sedimentary rocks that is insoluble in nonoxidizing acids, and organic solvents. Kerogen in rocks has four principal sources: marine and **lacustrine**, **terrestrial**, and recycled. Most of the world's oil has formed from marine and lacustrine kerogen, whereas most coal is from terrestrial plants, and the recycled kerogen is largely inert. The relative ability of a source rock to generate petroleum is defined by its kerogen quantity (TOC) and quality (high or low in hydrogen). Whether or not it has generated petroleum is defined by its state of **maturation** (immature, mature, or postmature with respect to oil).

The quantity of organic matter usually is expressed as total organic carbon (TOC). The color of a rock is a rough but not always reliable indicator of its TOC content. Most sandstones and red beds have very low TOCs because the organic matter has been destroyed by oxidation. TOCs generally increase in shales as the color goes from red to **variegated**, to green, gray, and finally to black. Baker (1972) made a detailed study of the Cherokee Shales (Pennsylvanian age) of Kansas and Oklahoma. He found TOCs as low as 0.1% in greenish gray shales and as high as 17.5% in one black shale unit.

The quantity of TOC in rocks is closely related to sediment particle size. A sample of Viking Shale from Alberta, Canada, was disaggregated, **dispersed** in water, separated by **centrifuging**, and analyzed for TOC. The TOC content of the **siltstone** size was 1.47%; clay size 2μm to 4μm, 1.70%; and **clay** size less than 2μm, 5.32%. A similar study of carbonates showed the highest concentration of TOC in the **lime** muds and the lowest values in the **skeletal** grains. High TOCs in sediments are due to the preservation and transport of organic matter, not organic productivity. Lowest in preservation are the high-energy parts of coastal areas and inland seas where productivity in the water column is adequate, but strong currents and the **chemical degradation** of the TOC. The preservation of TOC generally increases from high-energy to low-energy sediments. Inland seas such as the Caspian and silled basins such as Lake

Maracaibo both show their highest TOCs in the fine-grained sediments and their lowest TOCs in the coarse sediments.

Low-energy coastal areas and inland sedimentary basins where fine-grained clay and carbonate muds are deposited generally contain 0.5% to 5% TOC, which is in the range of most oil-forming rocks. Shallow inland seas, narrow seaways between continents, and restricted areas are the typical depositional environments for source beds of petroleum.

Even larger quantities of organic matter are preserved in areas where oxygen is eliminated and **benthic** organisms are suppressed. Sediments with TOC contents exceeding 10% are found in **stagnant**, silled basins, like the Norwegian **fjords** and the Black Sea, where **hydrogen sulfide** in the bottom water eliminates all life except **anaerobes**, such as **sulfate** reducers. The lack of oxygen restricts decomposition to reducing processes, and the poisonous effect of hydrogen sulfide kills all **biota** venturing into the area.

All **living matter** is composed of four main constituents: **lipids**, proteins, **carbohydrates** and **lignins**. Lipids cover all organism-produced substances that are practically insoluble in water. This encompasses fatty substances such as vegetable oil, waxes and animal fats. Lipids are structurally very similar to the non-aromatic NSO compounds. Lipids are one of the main sources for petroleum.

Proteins are high-order **polymers** made from individual **amino** acids, and account for most of the nitrogen and sulfur compounds in organisms. Therefore, most nitrogen and sulfur containing compounds found in petroleum are derived from proteins.

Carbohydrate is a collective name for individual sugars and their polymers. The name is derived from the formula $C_n(H_2O)_n$, which suggests hydrated carbons. Carbohydrates are not important sources for petroleum.

Lignins are widespread in plants and are characterized by their aromatic (phenolic) structures. They are high molecular weight structures (polyphenols). The aromatic content of petroleum is derived from this group.

Lipids are found in the lower life forms, such as algae, plankton and lower plant forms, and that lignins are found only in high-order terrestrial plants (due to its stiff, supportive structure). The main contributors to organic matter in sediments are: **phytoplankton**, **zooplankton**, higher plants, and bacteria.

There are three major types of kerogen, each dependent on the type of source material. Type Ⅰ kerogen is characterized by having a high initial hydrogen to carbon atomic ratio (H/C) of 1.5 or more, and a low oxygen to carbon atomic ratio (O/C) of less than 0.1. Its primary source is from algal sediments, such as lacustrine deposits. Type Ⅰ kerogen is also called alginite kerogen, containing high concentrations of alkanes and fatty acids. It is the best source for oil-prone maturation, but unfortunately it is very rare.

Type Ⅱ kerogen has a relatively high H/C ratio (1.0 to 1.4) and a low O/C ratio (0.09 to 1.5). It consists of abundant moderate length aliphatic chains and naphthenic rings. Ester bonds are common and sulfur is present in substantial amounts. Type Ⅱ kerogen is also called **exinite**, and is usually associated with marine sediments, where autochthonous organic matter (bacteria, phytoplankton and zooplankton) has been deposited in a reducing environment. It is a good oil or gas prone kerogen. It is more common than alginite.

Type Ⅲ kerogen has a relatively low H/C ratio (usually less than 1.0) and low O/C ratio (0.2 to 0.3). It contains an important proportion of polyaromatic nuclei and heteroatomic ketone and carboxylic acid groups. Aliphatic groups are a minor constituent, usually consisting of longer chains originating from higher-order plant waxes. The main source of this type of kerogen are continental plants found in thick detrital sedimentation along continental margins. This type of kerogen is also called vitrinite. It is less favorable for oil generation, but will provide a source rock for gas.

All kerogen types experience chemical alteration during maturation. This results in the formation of petroleum, and generally begins with the loss of oxygen, followed by hydrogen, to arrive at a form of hydrocarbon. The terms "sapropelic" and "humic" organic matter are often referred to when using visual descriptions of kerogen. Humic material is thought to be derived from plant matter, while sapropelic material originates from algae or plankton.

New Words and Phrases

1. immature *adj.* 未成熟的
2. erosion *n.* 剥蚀，侵蚀
3. shale *n.* [地]页岩
4. lacustrine *adj.* 湖泊的，湖相的
5. terrestrial *adj.* 陆地的，陆相的
6. maturation *n.* 成熟，成熟作用
7. variegated *adj.* 杂色的，多样化的
8. dispersed *adj.* 分散的，散布的
9. centrifuging *n.* 离心法，离心过滤
10. siltstone *n.* [地]粉砂岩
11. clay *n.* 黏土
12. lime *n.* 石灰，灰泥
13. skeletal *adj.* 骨骼的
14. chemical degradation 化学降解
15. benthic *adj.* 深海底的
16. stagnant *adj.* 停滞的
17. fjord *n.* 海湾，峡湾
18. hydrogen sulfide 硫化氢
19. anaerobic *adj.* [微]厌氧性的
20. sulfate *n.* [化] 硫酸盐
21. biota *n.* [生态]生物区
22. living matter 生物质
23. lipid *n.* 脂质，类脂
24. carbohydrate *n.* [化]碳水化合物
25. lignin *n.* [生化]木质素
26. polymer *n.* 聚合体
27. amino *adj.* [化]氨基的
28. phytoplankton *n.* 浮游植物
29. zooplankton *n.* 浮游动物
30. exinite *n.* 壳质组

Lesson 24

Lesson 24 ▶ Reservoir Rock

Oil and/or gas **reservoir rock** is understood as not only a rock that can contain but also yield these fluids during the development and production in a given temperature-pressure and geochemical (**physicochemical**) environment. Physical and chemical conditions may cause alterations, such as irreversible precipitation of salts out of solution (water, oil, gas) or, conversely, dissolution of salts, etc. Exposure to air (oxidation) also changes the **wettability** of rocks (oil-wet versus water-wet). Logs can provide continuous record of petrophysical properties, which can be compared to those obtained by core analysis. **Seismic** surveys enable one to determine some rock **parameters** even without drilling the wells.

Important characteristics of reservoir rocks are their capacity to contain certain volume of oil and/or gas and their capacity to yield oil and/or gas. The first characteristic is defined by **porosity** and the second one by **permeability**.

The total volume of void space in the rock, including pores, **vugs**, and fractures, is called the total or absolute porosity. Total porosity is the ratio of the total void volume to the bulk volume. It is expressed as a fraction or percentage.

Some pores are not interconnected. Such isolated pores are not involved in the fluid flow during the development and production. Some pores and channels do not permit fluid movement and turn out to be ineffective due to their small diameter, wettability of the channel walls, and irreducible **fluid saturation**. Thus, the ratio of the effective pore volume to the bulk volume is called effective porosity (fraction or percentage). Effective porosity must be always determined for a specific fluid and for the specific reservoir conditions. It is equal to open porosity minus the irreducible fluid saturation. It may be determined by means of **petrophysics** (logs) or using special field studies.

Porosity depends first of all on the **grain size**, packing, sorting, roundness, and the mineralogy and amount of cement. Porosity also depends on the occurrence and preservation of vugs and fractures formed as a result of secondary alterations.

Structure and texture of rocks strongly affect the geometry of the **pore space**. Structure of rocks means external features of rock grains (their shape, nature of surface, etc.). Texture includes the type of interrelations between the grains and their orientation. In particular, **lamination** is the most important and commonly used texture feature. Texture of the sedimentary rocks is born during sedimentation. Although post-depositional alterations can significantly affect the texture, recognizable features of the initial texture are usually preserved.

A significant influence on the interrelation between the rocks and fluids is the specific surface area. In **clastic** rocks, the specific pore surface area is inversely proportional to the grain size as follows: $S_p = 6(1-\phi)d$, where S_p is the specific pore surface area (cm^2/cm^3 or cm^{-1}), ϕ is the porosity (fractional), and d is the average grain size (cm).

The density of sedimentary rocks ranges from $1.7g/cm^3$ to $2.6g/cm^3$. In clastic rocks, density is inversely proportional to porosity. Carbonate rocks often form oil and gas reservoirs. **Reef** buildups, **calcarenites**, and **oolitic** limestones all have **primary porosity**. This primary porosity substantially changes during **diagenesis** upon recrystallization, **dolomitization**, and leaching. The latter process is very important in the karst formation. The formation of **karst** can begin at the stage of early diagenesis and then continue during **epigenesis** (catagenesis), especially within highly **fractured** rocks. Unfortunately, vugs in vuggy limestones are often filled by the second - generation **calcite** and other newly formed minerals. Dolomitization processes can increase porosity up to 13.1%, whereas sulfatization and **silicification** can significantly decrease porosity. Porosity due to fractures alone is around 1%, whereas in the presence of vugs and cavities, porosity can reach 3.5%.

Permeability is the ease of flow of fluids through rocks. The volumetric rate of flow is proportional to the pressure gradient: $Q=KAD_p/\mu L$, where Q is the flow rate (m^3/s), μ is the **dynamic viscosity** (Pa · s), D_p is the pressure gradient along the length L (in m, Pa/m), A is the cross-sectional area (m^2), and K is the permeability (m^2).

Permeability is a function of the pore size and shape, pore throat and/or channel diameter, grain size and shape, grain **packing** density, tortuosity, sorting, cementing, fracturing, and residual fluid saturation. The above definition of permeability indicates that its value should not be affected by the nature of a liquid moving through the **porous medium**. Actually, however, permeability changes depend on the type of flowing fluid. These changes are sometimes greater than 100% . According to definition, permeability also should not change with time. Experiments, however, often demonstrate permeability drop of up to 50% within 1 h. There are different explanations of the reasons causing permeability change in time, and the effect of fluid properties on permeability. If fluids flow through loose reservoir rocks that include some fine sands, rock grains may change their positions, and the pore channels may become **plugged** with fine material. Colloidal particles suspended in oil may precipitate and plug the pores. Resins and asphaltenes present in crude oil may also precipitate, and result in a decrease of the cross-sectional area of pore channels, throats, and canals. Wettability of rocks (oil - wet versus water-wet) also changes the relative permeability to water and to oil.

When water flows through reservoir rocks that include clay minerals, many of the clay minerals swell, which also results in a decrease in the cross-sectional area of pore channels. Water in contact with silica may give rise to colloidal silica in porous space, which may lead to plugging of the pore channels. When CO_2 is released from water, $CaCO_3$ precipitates within reservoirs, decreasing the pore throat and canal diameters.

Beside clastic and carbonate rocks, reservoirs may be composed of volcanic rocks, volcaniclastic rocks, and shales. Oil accumulations in such reservoirs are found in many countries (Azerbaijan, Turkey, Cuba, USA, Indonesia, etc.)

Reservoir properties of rocks are most commonly encountered in the ultrabasic, basic, and medium extrusive rocks, and rarely, in rhyolites. Volcaniclastic and mixed volcanic - sedimentary reservoirs are quite common. Effective porosity in such reservoirs is sometimes associated with intergranular pores and vugs, but, as a rule, with fracturing. An important role in the formation of reservoirs included in this group belongs to the fractures formed as a result of weathering and leaching (in particular, hydrothermal leaching). High-quality reservoirs in such sequences, however, are sporadic. One may encounter very diverse oil flow rates in the adjacent wells, from almost nil in one well to a few thousand tons a day in the next one.

New Words and Phrases

1. reservoir rock 储层
2. physicochemical *adj.* 物理化学的
3. wettability *n.* 润湿性
4. seismic *adj.* [地]地震的
5. parametric *adj.* [数][物][晶]参(变)数的，参(变)量的
6. porosity *n.* 多孔性，孔性
7. permeability *n.* 渗透性，渗透率
8. vug *n.* 孔洞
9. fluid saturation [地质] 流体饱和度
10. saturation *n.* 饱和(状态)，浸润，浸透饱和度
11. petrophysics *n.* 岩石物理学
12. grain size 粒度，颗粒尺寸，结晶粒度(大小)
13. pore space 孔隙，孔隙空间
14. lamination *n.* 层理
15. clastic *adj.* [地质]碎屑状的
16. reef *n.* 生物礁
17. calcarenite *n.* 砂屑灰岩
18. oolitic *adj.* [地]鲕粒岩的
19. primary porosity 原生孔隙
20. diagenesis *n.* [地质]成岩作用
21. dolomitization *n.* 白云石化作用
22. karst *n.* 喀斯特，岩溶
23. epigenesis *n.* 晚期成岩作用
24. fractured *adj.* 断裂的
25. calcite *n.* [矿]方解石
26. silicification *n.* [化]硅化(作用)
27. dynamic viscosity 动力黏度
28. packing *n.* 堆积
29. porous medium 多孔介质
30. plugged *adj.* 被堵塞的

Lesson 25 ▸ Seal Rock

In order for petroleum to be trapped in the subsurface, impermeable rocks of some sort need to be present to keep oil and gas in the trap and prevent the upward migration of these hydrocarbons to the surface. The impermeable rocks that fulfill this function are known as seal rock.

Seal rock is a formation with extremely low porosity and permeability overlying an oil or gas reservoir, and it constitutes the barrier against the volume flow of hydrocarbons into the upper layers. Although a seal rock can be considered as a seal to hydrocarbons, it is erroneous to regard it as a completely impermeable layer. Two main mechanisms have been well recognized to be responsible for migration of hydrocarbon gases through seal rocks into adjacent upper layers. One is molecular diffusion through the water-saturated pore space of the seal rock. The other is the compressible slow Darcy flow of a free gas phase. Molecular diffusion is a ubiquitous but slow process that is only considered significant in geological timescales. The second one, the slow Darcy flow, depends strongly on the geologic and hydrodynamic conditions of the system, including the reservoir, the seal rock, and the overburden formations, as well as the properties of the fluids in both the reservoir and the seal rock. Slow Darcy flow occurs when the pressure difference across the seal rock is sufficiently high to overcome the sealing capacity of the seal rock. In essence, the sealing capacity of a seal rock is provided by the **capillary** forces across the interface of the wetting phase (usually brine), which saturates the seal rock, and the nonwetting phase (oil or gas), which accumulates in the reservoir. The capillary sealing mechanism is illustrated schematically in Figure 25. 1.

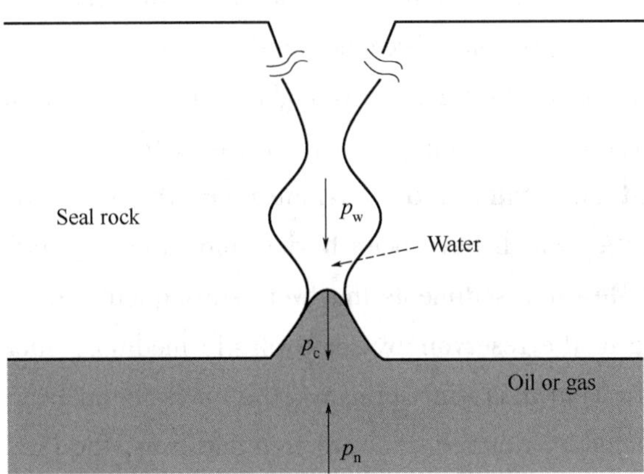

Figure 25. 1 Schematic of capillary sealing mechanism in a pore throat of seal rock

Figure 25. 1 shows a pore throat and a curved interface between the wetting and nonwetting phases in a seal rock, where p_n is the pressure in the nonwetting phase, p_w is the pressure in the wetting phase, and p_c is the capillary pressure across the nonwetting/wetting meniscus in a pore throat. It is the capillary pressure that prevents the penetration of the nonwetting fluid into the seal rock through slow Darcy flow. When the pressure difference between the nonwetting and the wetting phase exceeds the capillary pressure at a pore throat, i. e. $p_n - p_w > p_c$, the nonwetting phase will advance along the channel until it reaches the next smaller pore throat. When the differential pressure across the seal rock overcomes the capillary pressures of a series of interconnected pore throats of arbitrarily large sizes, a continuous filament of nonwetting phase will be formed and, consequently, a slow Darcy flow will occur. This differential pressure is regarded as the breakthrough pressure of the seal rock.

The breakthrough pressure is an important parameter for assessing the sealing capacity of a seal rock of a hydrocarbon trap. The magnitude of the breakthrough pressure is determined by the highest capillary pressure of an interconnected network of pore throats that are first invaded by the nonwetting phase.

Shale, salt, and cemented sandstones are all potential seals. However, as oil matures in the reservoir and becomes lighter and more buoyant due to the effects of temperature, it may eventually be able to overcome the effectiveness of a seal and start to leak out of the reservoir. Thus, rocks that were once great seals will start to leak when the day arrives that the buoyancy of the hydrocarbons in the trap exceeds the capillary entry pressure of the seal. However, once the buoyant hydrocarbons have escaped a seal may once again be able to trap hydrocarbons and create a pool of oil or gas.

It will be appreciated that it is insufficient for oil to have been formed and to have migrated under suitable conditions into a nearby **structural** or **lithological trap** for an oilfield to be found in that locality today.

It is also necessary for the oil **reservoir**, whether it be a structural or lithologic one, to have been closed after the accumulation of the oil and protected against **subsequent** Earth movements which would otherwise have breached and drained the reservoir of its fluids.

This essential process of sealing may be brought about in the following ways:

(1) By the relatively rapid accumulation of a **mass** of thick sediments over the reservoir, which are thus **shielded** and **blanketed**. Examples are the Palaeozoic fields of the Mid-continent basin of the USA, which have been buried and protected and therefore preserved by the great thicknesses of Mesozoic sediments that were subsequently deposited above them.

(2) By the sealing of the reservoir by some plastic medium, such as **salt**, **gypsum**, or certain types of shale. In Thuringia, in Germany, the oil is found below a thickness of 1,200 feet of salt, which has acted as a perfect seal. In Iraq and Iran, the Pars series, which contains plastic shales, **anhydrite** and salt layers, form an excellent cover which has proved adequate protection against all but the most violent shocks. At Boryslaw (Poland), the Miocene plastic clays act as cover for the overthrust folds of the Oligocene Boryslaw sandstone, although some fissuring at the apices has allowed a portion of the oil to escape and form **ozokerite** veins. When

plastic beds are present, they will act as a lubricating medium, allowing the beds above to take on secondary folding, so that the underlying structures may have no relation to the surface structures (e. g. the Iran fields, where the overlying Bakhtiari and Pars structures are very different from the underlying Asmari limestone structure).

(3) By the formation of a hard, impervious cap rock over the oil reservoir. This is a very common **feature** of oilfields, and the cap rock usually consists of an impervious shale or a dense limestone.

It has been shown experimentally by Lockwood that oil and gas cannot accumulate beneath an impervious cap rock if the reservoir sand is water–saturated to begin with. Hence, the rock immediately above an oil reservoir must originally have been permeable enough to let compaction currents pass through it while the oil was accumulating. How, then, has it subsequently become **impervious**? The chief cause appears to lie in the adhesive properties of the oil, which are increased by the chemical action resulting from the bacterial reduction of sulphates. Taylor′s theory is that shales in which base exchange has taken place not only provide a satisfactory cover rock for oil deposits when formed, but also provide the **alkaline anaerobic** conditions which favour the formation of hydrocarbons.

New Words and Phrases

1. capillary *adj.* 毛(发)状的, 细长的, 毛细管作用(现象)的
2. structural *adj.* 结构的, 建筑的
3. lithologic *adj.* 岩性的
4. trap *n.* 圈闭, 陷阱
5. reservoir *n.* 储层, 油藏
6. subsequent *adj.* 后来的, 随后的
7. mass *n.* 块, 团; *adj.* 大规模的, 集中的
8. shield *n.* 盾, 防护物
9. blanket *n.* 毯子, 毯状物, 覆盖层
10. salt *n.* 盐, 风趣, 刺激性
11. gypsum *n.* 石膏, 石膏肥料
12. anhydrite *n.* 硬石膏, 无水石膏
13. feature *n.* 容貌, 特色, 特征
14. alkaline *adj.* 碱性的, 碱的
15. anaerobic *adj.* 厌氧的
16. impervious *adj.* 不可渗透的, 透不过的
17. ozokerite *n.* 地蜡, 石蜡

Lesson 26

Lesson 26 ▸ Traps

A traps is a geological feature which enables migrating petroleum to accumulate and be preserved for a certain time interval. Traps occur in fundamentally different forms and can enclose very different volumes of pore space and hence petroleum. The maximum total holding capacity or closed volume of a trap is the volume between its highest point and the "spilling plane" or outflow level at the bottom. Traps are rarely completely full, i. e., the **petroleum column** rarely extends down to the point at which it would spill out of the trap. Furthermore, traps are never full in the sense that all available pore space in the reservoir rock is occupied by petroleum. There is always a certain residual amount of water, which can not be displaced by migrating petroleum.

Traps can be formed either by tectonic activity such as faults, folds, etc. (**structural traps**) or from sedimentary depositional patterns (**stratigraphic traps**). The majority of known oil fields is found in structural traps, but structural traps are easier to locate by current geophysical and geological exploration techniques. Stratigraphic traps have become a prime target for exploration only in more recent years. Sometimes there are combined tectonic and depositional causes for the formation of a trap. Then these traps are called **combination traps**.

Anticlinal structures are the most common type of traps. One of the largest oil-producing **anticlines** is the Kirkuk field in Iraq. The productive area extends in three culminations over a length of nearly 30 miles. The "Golden Lane" fields in Mexico also occur as local "highs" along an arch extending some 50 miles. Most oil-producing anticlines, however, are only a few miles in length, and are often associated with some faulting.

The giant Wilmington oil field in the Los Angeles basin of southern California may be cited as an example of a structural anticlinal trap. The Los Angeles basin is one of the most prolific oil-producing basins in the world in terms of the ratio of pooled oil to total rock volume. The Wilmington structure is a broad, **asymmetric anticline** about 12 miles long and 4 miles wide. It is dissected by a series of **transverse normal faults** which separate the producing sand and sandstone reservoirs into many different pools. **Seals** are provided by claystones and shales. Oil properties in the various **fault blocks** differ considerably, illustrating that the accumulation of hydrocarbons apparently occurred independently in each of the blocks. The different oil properties also suggest that the faults provide effective barriers to communication between adjacent fault blocks. There are seven major producing zones ranging in age from late **Miocene** to early **Pliocene**. The upper part of the Wilmington structure was eroded during the early Pliocene, and subsequent submergence resulted in the deposition of an additional 1800~2000 ft of almost horizontal Pliocene and Pleistocene sediments on top of the anticline. The entire Wilmington oil field is estimated to contain about 3 billion barrels of producible oil.

A different type of structural trap is associated with so-called **growth faults** which play an important role in such prolific areas as the Niger Delta region and the Gulf Coast, USA. The term growth fault is used because the faults remain active after their formation, and allow faster sedimentation on the down-thrown side. The enhanced deposition of sediments across the growth fault initiates a rotational movement which tilts the beds toward the fault—creating anticlinal rollover structures along the faults.

There are numerous oil fields where oil is trapped in monoclines against seals provided by fault planes above the **monocline**. The Pechelbronn oil field of the Rhine Graben and the oil fields in **Cretaceous** Woodbine sandstones along the fault zone of Mexia in eastern Texas are examples of such fault traps. It should be pointed out again, however, that **fault zones** are not always sealing, but are sometimes nonsealing. Fault zones may even provide **migration avenues**, as in the Niger **Delta** or other areas with deep reaching fault systems.

The association of oil accumulations with salt domes is another example of structural traps formed by tectonic events. Oil accumulations in tectonic features caused by the upward movement of salt from deeply buried evaporate beds are known in several parts of the world, including northern Germany and the North Sea, Romania, West Africa and the Gulf Coast region of USA and Mexico. The relationship between the oil accumulations and the salt is only structural and there are many ways to provide suitable traps in association with salt intrusion. The traps may be formed along the folded or faulted flanks of the **salt plug** or on top of the plug where arching and/or faulting is produced in the overlying sediments.

The formation of stratigraphic traps is related to sediment deposition or erosion and is thus distinguished from formation of structural traps, which originate from tectonic events. It is clear, however, that **tectonism** ultimately controls many geological processes such as deposition and erosion of sediments. Stratigraphic traps can be related to facies changes, **diagenesis** of sediments and to **unconformities** in the **sedimentary rock column** . Most basins contain the prerequisites for the creation of stratigraphic traps. Typical examples are barrier sand bars, **deltaic distributary channel sandstones** and **carbonate reefs** of different forms and sizes. Such barrier and channel sands and carbonate reef structures, frequently embedded in fine-grained organic-rich sedimentary sequences, are ideal traps for migrating petroleum.

Fossil organic reefs form very prolific-oil stratigraphic traps in many parts of the world, for example, the Middle East, Libya, southwestern USA and in Alberta, Western Canada. Reefs are constructed by various groups of organisms, such as corals, algae, **echinoderms**, **mollusks** and foraminifera. They may be interpreted as "fringing reefs", "barrier reefs", "atolls" or "pinnacle reefs", depending on their environment of growth and shape. Accordingly they vary in length, width and thickness. As an example of reef traps the giant oil fields of the Middle Devonian reefs of the Rainbow area in Alberta, are briefly discussed from a paper by Barss et al. The growth of the Rainbow Member reefs in the Black Creek Basin of northwestern Alberta was initiated during Upper Keg River time when banks of reef-constructing organisms started to flourish. Structural control of reef growth, if it did exist, was subtle. The reefs in the Rainbow sub-basin, atolls and pinnacle reefs, have vertical reliefs up to 820 ft. The reefs are enveloped by fine-grained off-reef sediments, consisting of salt, interbedded anhydrites, dolomites and shales. An evaporate cover provides an excellent seal for **petroleum entrapment**. Reserves from the Rainbow Member pools are estimated to be in excess of 1.3 billion barrel of oil-in-place.

Finally, it should be pointed out that migrating petroleum can be captured by all kinds of traps, by structural, stratigraphic or combination traps, only as long as they are in the path of migration.

New Words and Phrases

1. anticline *n.* 背斜
2. seal *n.* 盖层
3. Miocene *n.* & *adj.* 中新世(的)
4. Pliocene *n.* & *adj.* 上新世(的)
5. Cretaceous *n.* & *adj.* 白垩纪(的)
6. delta *n.* 三角洲
7. monocline *n.* 单斜(层)
8. tectonism *n.* 构造作用
9. diagenesis *n.* 成岩作用
10. unconformity *n.* 不整合
11. echinoderm *n.* 棘皮动物
12. mollusk *n.* 软体动物
13. recoverable reserves 可采储量
14. asymmetric anticline 不对称背斜
15. transverse normal faults 横向正断层
16. petroleum entrapment 石油圈闭
17. growth faults 生长断层
18. fault zones 断裂带
19. spill point 溢出点
20. petroleum column 油柱
21. stratigraphic trap 地层圈闭
22. structural trap 构造圈闭
23. combination trap 复合圈闭
24. fault blocks 断块(构造)
25. migration avenues 运移通道
26. salt plug 盐栓(盐丘)
27. sedimentary rock column 沉积岩剖面
28. carbonate reefs 碳酸盐礁
29. deltaic distributary channel sandstones 三角洲分流河道砂岩

Lesson 27 ▸ Migration and Accumulation

Once petroleum has reached the reservoir rock, it must be **concentrated** into pools if it is to be commercially available. As we have seen, it conceivably might have entered the reservoir rock as organic matter, possibly in part water-soluble, possibly colloidal, but most likely it was already in the form of petroleum or a petroleum-like substance. It may have been squeeze directly out of the **muds** and **clays**, it may have remained until the clay became shale and then been **squeezed** out of the **shale**, or it may have been **deposited** as petroleum **hydrocarbons** directly in the reservoir rock. At some time in its history, however, the source material was transformed into petroleum. And this petroleum, as either oil or gas or as both, got into the reservoir rock, where it was ready to accumulate into pools. This final movement into pools may be called the **secondary migration**, as distinguished from the primary migration, which brought the petroleum from the source into the reservoir rock. Again we must admit, as in discussing the origin and the **primary migration** of oil and gas, that we do not have a completely satisfactory answer to the problem of the movement of oil and gas within the reservoir rock and their concentration into pools. We know where to look for petroleum, but we do not know how it got there.

Presumably the petroleum that entered the water-**saturated** reservoir rocks was dispersed in minute particles, possibly of **colloidal** size, and some might even be in solution in the water. In the absence of any unbalanced forces, such petroleum might remain more or less stationary for a long time and might be **buried** deep. Local pressure or temperature **gradients** might cause local movements, but it would require some regional change to **upset** the **equilibrium** and cause a regional movement. This might be regional folding, tilting, mountain-building, or warming up—possibly due to **igneous** activity or to various changes in the **hydrodynamic** gradients. The normal geologic history of most sedimentary regions includes many such episodes that may have upset the equilibrium of the reservoir fluids, and each may have caused some increment of movement.

Before we discuss the theories that attempt to explain these movements, it will be useful to point out several fundamental aspects of the problem that should be kept in mind. (1) The geological conditions or frame within which petroleum migration and accumulation must have occurred. (2) The distances through which petroleum has migrated. Are we limited to short distances of migration—say less than a mile? (3) How much of the migration was **vertical**, and how much of it was **lateral**?

Geological Framework of Migration and Accumulation

From knowledge of the geological conditions that prevail in the oil-producing regions, it is possible to set up a framework of limiting geological factors within which any theory of petroleum migration and accumulation must fit. Such a frame can at best be only general, for it includes many variables, both known and unknown.

The sides of the frame might be summarized as follows: the environment's hydrodynamics, the reservoir rocks and the trap, the time of oil and gas to originate, migrate, and accumulate into pools, the upper boundary or roof rock of every pool, the temperatures and the pressures of the reservoir, the geologic history of the trap.

Short or Long Migration

One of the fundamental questions involved in the accumulation of petroleum is whether petroleum migrates considerable distances (i. e. , one or more miles) to form pools, or whether it was formed in place. Some believe that migration was negligible and that petroleum was formed virtually where it is now found. If this is so, there is no problem of secondary migration, and exploration for petroleum should be confined chiefly to the search for favorable traps in areas of origin.

In some pools there is good geological evidence that the oil and gas were formed near by or even in place. The distance implied by the term "short-distance migration" is arbitrarily taken as not more than a mile.

There are good reasons, on the other hand, for believing that some oil and gas may have migrated for long distances to accumulate into pools. Some one believes that the oil was squeezed out of the sediments in the mobile and deformed belts of the Earth, and entered the water-bearing layers, which calls "carrier beds", in which it was carried to pre-existing traps.

We may conclude, therefore, that in some pools there is good evidence that the petroleum was formed in place or migrated only a short distance, while there is equally good evidence that the oil in other pools migrated for a long distance. Obviously, if petroleum will migrate a distance measured in feet, it can, by increments, multiply the distance into miles and tens or even hundreds of miles during geologic time. The distance it may migrate is determined by the distance from the source area to the nearest trap; if this distance is short, as it usually is, the distance the petroleum has moved is short; but, if there is no obstruction to migration through the reservoir rock, we may expect the petroleum to keep on moving until it gets to a trap or until it is lost through escape at the surface. It may have to move tens or even hundreds of miles, but the chances of its having to go so far without finding a trap are slight.

Vertical Compared with Lateral Migration

The extent of vertical migration of oil and gas, as compared with lateral migration, has received considerable thought. In some fields a number of pools lie one above another. In speculations about the source of the oil in the shallower pools, it frequently seems more natural to look to the deeper oil-bearing formations, from which oil might have migrated vertically

upward along **fractures**, **faults**, and **fissures**, than to lateral migration. The movement of oil in the multi-sand fields of the Baku province, USSR, for example, is thought by some to have been along vertical fractures and faults. Some of the considerations that may apply in a situation where one pool overlies another or others are these:

(1) Where the reservoir pressures vary erratically from reservoir to reservoir, it seems unlikely that there are any vertical connections; interconnections seem probable, on the other hand, when the pressures differ uniformly by the weight of the intervening column of water.

(2) The presence of intervening water sands between formations producing oil and gas seems good evidence that vertical movement of the petroleum has not occurred. How could the oil pass through the water sand without leaving some behind or pass through some water sands and be retained by others?

(3) Differences in oil analyses from different sands would indicate that there is no present **interconnection**, whereas similar oil analyses might mean either vertical or lateral movement.

(4) **Erratic** gas-oil ratios would be evidence against interconnection, especially if the lower producing formations contained free gas and the **shallower** sands were under-saturated.

(5) The fact that many traps are **sealed** by faults suggests that faults are not to be considered generally as avenues of oil migration. However, this may not have been the situation at all times in the past. Faults and fractures may have permitted vertical movements for a while and then become sealed, and other changes subsequent to the formation of the pools may have altered the composition of the fluids and changed the pressure relations.

(6) Sandstone **dikes**, which are present in some areas, may act as conduits for the vertical migration of oil and gas. Some of them appear to have done so, for they are saturated with oil where they crop out.

We may conclude that petroleum may have moved vertically along fault planes and fracture openings in some fields, but most of the evidence indicates that most petroleum probably arrived at the trap by lateral migration. The theory of oil-finding in America is based chiefly on lateral migration; faulting, if present, is considered more likely to be a part of the trap, keeping the oil in, than a means of escape permitting migration into shallower formations.

New Words and Phrases

1. migration *n.* 移动，迁移
2. accumulation *n.* 积聚，累积，堆积物
3. concentrate *vi.* & *vt.* 聚集，集中，浓缩
4. mud *n.* 泥；泥浆
5. colloidal *adj.* 胶质的，胶体的，胶状的
6. clay *n.* 黏土
7. squeeze *vt.* & *vi.* 挤，勒索，压榨
8. shale *n.* 页岩，泥板岩
9. deposit *n.* & *vi.* 沉积，沉淀，矿床
10. hydrocarbon *n.* 碳氢化合物
11. secondary migration 二次运移
12. primary migration 初次运移

13. saturate *vt.* & *adj.* 饱和(的)，浸透(的)
14. bury *vt.* 埋葬，隐藏
15. gradient *n.* 梯度，倾斜度
16. upset *vt.* 翻倒，颠覆
17. equilibrium *n.* 均衡，平静
18. igneous *adj.* 火(成)的
19. hydrodynamic *adj.* 水动力学的，水力的
20. vertical *adj.* 垂直的
21. lateral *adj.* 侧面的，横向的
22. upward *adj.* 向上的，上升的
23. fracture *n.* & *vt.* 破裂，断裂
24. fault *vi.* 错断，产生断层
25. fissure *n.* & *vi.* 裂缝，裂沟，裂开，分裂
26. interconnection *n.* 互相联络
27. erratic *adj.* 不稳定的
28. shallow *adj.* 浅的；*n.* 浅滩
29. sealed *adj.* 封口的，密封的
30. dike *n.* 堤防

Module 3

Exploration of Hydrocarbon Fields

03

Lesson 28 ▸ Sedimentary Basins

Lesson 28

A sedimentary basin is an area of the Earth's crust that is underlain by a thick sequence of sedimentary rocks. Hydrocarbons commonly occur in sedimentary basins and are absent from intervening areas of igneous and metamorphic rocks. This fundamental truth is one of the cornerstones of the sedimentary-organic theory for the origin of hydrocarbons. Therefore it is important to direct our attention not only to the details of traps and reservoir rocks but also to the broader aspects of basin analysis. Before acquiring **acreage** in a new area and long before attempting to locate drillable **prospects**, it is necessary to establish the type of basin to be evaluated and to consider what productive **fairways** it may contain and where they may be extensively located. This section describes the various types of basin with reference to examples from around the world and discusses the relationship between the genesis and evolution of a basin and its **hydrocarbon potential**.

First, however, some of the basic terms and concepts must be defined. A sedimentary basin is an area on the Earth's surface where sediments have accumulated to a greater thickness than they have in adjacent areas. No clear boundary exists between the lower size limit of basin and the upper limit of a syncline. More geologists would probably take the view that a length of more than 100km and a width of more than 10km would be a useful dividing line. Most sedimentary basins cover tens of thousands of square kilometers and may contain over five kilometers of **sedimentary fill**. Note that a sedimentary basin is defined as an area of thick sediment, with no reference to its topography. A sedimentary basin may occur as part of a mountain chain, beneath a continental **peneplain**, or in an ocean. Conversely, a present day ocean basin is not necessarily qualified as a sedimentary basin; indeed, many are floored by igneous rocks with only a veneer of sediment.

This distinction between topographic and sedimentary basin needs further elaboration. Both types of basin have a depressed basement. Sedimentary basins may or may not have been marked topographic basins during their history. Many basins are infilled with continental and shallow marine sediments, and totally lack deep-sea deposits.

Similarly, a distinction needs to be made between **syndepositional** and **postdepositional** basins (Figure 28.1). Most sedimentary basins indicate that subsidence and deposition took place simultaneously and their simultaneous occurrence is shown by facies changes and paleocurrents that are concordant with structure. On the other hand, in some basins paleocurrent directions and facies are discordant with and clearly predate present structure. This is particularly characteristic of **intracratonic** basins, as is shown later. The distinction between these two types of basin is critically important in petroleum exploration because of the need for traps to have formed before hydrocarbon generation and migration. Stratigraphic traps are generally formed before migration, except for rare diagenetic traps. Structural traps may predate or postdate migration, and establishing the chronology correctly is essential.

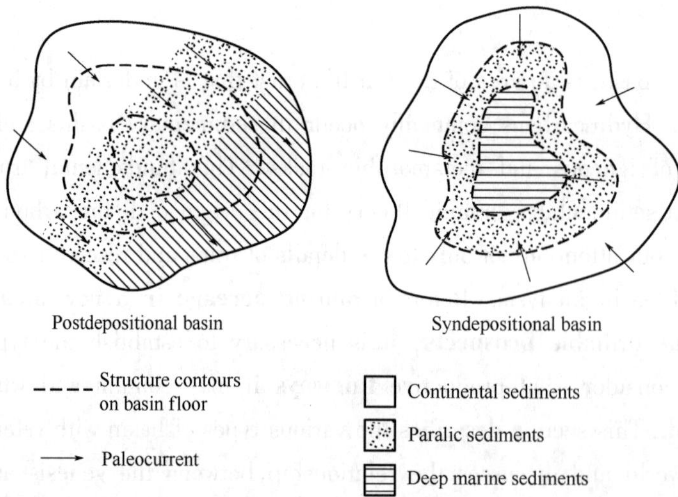

Figure 28.1 The differences between syndepositional and postdepositional sedimentary basins (Selley, 1998)

A further important distinction must be made between topography and sediment thickness. When examining regional **isopach** or **isochron** maps, it is tempting to assume that they are an indication of the **paleotopography** of the basin. This is by no means always true. The **depocenter** (area of greatest sediment thickness) is not always found in the topographic nadir of the basin, but may frequently be a linear zone along the basin margin. This is true of terrigenous sediments, where maximum deposition may take place along the edge of a delta. Sediments thin out from the delta front both up the basin margin and also seawards. Similarly, in carbonate basins most deposition takes place along shelf margins, where organisms thrive in shallow, well-oxygenated conditions with abundant nutrients. Thus reefs, skeletal sands and oolite sands thin out toward basin margin sabkhas and basinward into condensed sequences of lime mud.

Many studies have shown that a depocenter may migrate across a basin. The topographic center of the basin does not necessarily move with it. Examples of this phenomenon have been documented from Gabon, Brazil and Iraq. Note that the thickness of each of the formations measured at outcrop should not be added together to determine the overall thickness of sediment within a basin. This measurement can only be made from drilling or geophysical data.

Now that basins have been considered in time and profile, they may be viewed in plan. The term basin has two interpretations. In the broadest sense, as already defined, a sedimentary basin is an area of the Earth's surface underlain by sediments. In a narrower sense basins may be subdivided into true basins; those that are subcircular in plan and those that are elongate.

Sedimentary basins form part of the Earth's crust, or lithosphere; they are generally distinguishable from granitic continental and basaltic oceanic crust by their lower densities and slower seismic velocities. Beneath these crustal elements is the more continuous **subcrustal** lithosphere. The crust is thin, dense, and topographically low across the ocean basins; but thick, of lower density, and consequently, higher elevation over the continents. The lithosphere is made up of a series of rigid plates, which overlies the denser, yet viscous, asthenosphere.

Basins can be formed in four main ways. One major group of basins, the rift basins, form as a direct result of crustal tension at the zones of sea floor spreading. A second major group of basins occur as a result of crustal compression at convergent plate boundaries. A third type of basin can form in response not to lateral forces but to vertical crustal movements. For reasons not fully understood, phase changes can take place beneath the lithosphere. These changes may take the form of localized cooling, and therefore contraction, resulting in a superficial hollow, which becomes infilled by sediment. Conversely, the lithosphere may locally heat up and expand, causing an arching of the crust. Erosion of this zone will then occur. Sometimes this crust doming is a precursor to rifting and drifting. Alternately, subsequent cooling and subsidence result in the formation of an intracratonic hollow, which may be infilled with sediments.

A fourth mechanism of basin formation is simple crustal loading due to sedimentation. This process poses a "chicken and egg" problem, however. Basins of this type require an initial depression in the crust before deposition may begin. Thus loaded basins characterize continental margins where a **prograding delta** can initiate and maintain the depression of adjacent oceanic crust.

New Words and Phrases

1. acreage *n.* 区块
2. prospect *n.* 远景圈闭
3. fairway *n.* 区带
4. hydrocarbon potential 油气潜力

5. sedimentary fill 沉积充填
6. peneplain *n.* 准平原，近似平原
7. syndepositional *adj.* 同沉积的
8. postdepositional *adj.* 沉积后的
9. intracratonic *adj.* 克拉通内的
10. isopach *n.* 等厚线
11. isochron *n.* 等时差线
12. paleotopography *n.* 古地形，古地形学
13. depocenter *n.* 沉积中心
14. subcrustal *adj.* 地壳之下的
15. prograding delta 前积三角洲

Lesson 29 ▸ Petroleum System

Lesson 29

A petroleum system is defined here as a natural system that encompasses a pod of active source rock and all related oil and gas and which includes all the geologic elements and processes that are essential if **a hydrocarbon accumulation** is to exist. This once-active source rock may now be inactive or spent (depleted). Petroleum here includes high concentrations of thermal or **biogenic gas** found in **conventional reservoirs** or in **gas hydrate**, **tight reservoirs**, fractured shale, and coal; or **condensates**, crude oils, and **asphalts** found in nature.

The term system describes the interdependent elements and processes that form the functional unit that creates hydrocarbon accumulations. The essential elements include a petroleum source rock, reservoir rock, seal rock, and **overburden rock**, and the processes are trap formation and the generation-migration-accumulation of petroleum. These essential elements and processes must occur in time and space so that organic matter included in a source rock can be converted to a petroleum accumulation. A petroleum system exists wherever the essential elements and processes occur.

Characteristics and Limits

The geographic, stratigraphic, and temporal extent of the petroleum system is specific and is best depicted using a table and the following four figures: (1) a burial history chart depicting the critical moment, age, and essential elements at a specified location; (2) a map; (3) a cross section drawn at the critical moment depicting the spatial relationship of the essential elements; (4) a petroleum system events chart (Figure 29.1) showing the temporal relationship of the essential elements and processes and the preservation time and critical moment for the system. The table lists all the oil and gas fields in the petroleum system.

The critical moment is that point in time selected by the investigator that best depicts the generation-migration-accumulation of most hydrocarbons in a petroleum system. A map or cross section drawn at the critical moment best shows the geographic and stratigraphic extent of the system. If properly constructed, the burial history chart shows that time when most of the petroleum in the system is generated and accumulating in its primary trap. For biogenic gas, the critical moment is related to low temperatures. Geologically, generation, migration, and accumulation of petroleum at one location usually occur over a short time span. When included with the burial history curve, the essential elements show the function of each rock unit.

The geographic extent of the petroleum system at the critical moment is defined by a line that circumscribes the pod of active source rock and includes all the discovered petroleum shows, **seeps**, and accumulations that originated from that pod. A plan map, drawn at the end

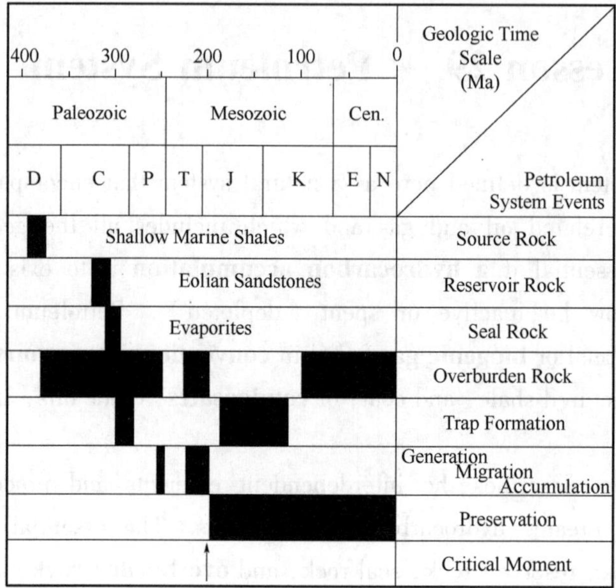

Figure 29. 1　Events chart for a petroleum system (modified from Magoon and Dow, 1994)

of Paleozoic time in our example, includes a line that circumscribes the pod of active source rock and all related discovered hydrocarbons. This map best depicts the geographic extent or known extent of the petroleum system.

Stratigraphically, the petroleum system includes the following rock units or essential elements within the geographic extent: a petroleum source rock, reservoir rock, seal rock, and overburden rock at the critical moment. The functions of the first three rock units are obvious. However, the function of the overburden rock is more subtle because, in addition to providing the overburden necessary to thermally mature the source rock, it can also have considerable impact on the geometry of the underlying migration path and trap. The cross section drawn to represent the end of the Paleozoic, shows the geometry of the essential elements at the time of hydrocarbon accumulation and best depicts the stratigraphic extent of the system.

The formation of traps is investigated using geophysical data and structural geologic analysis. The generation-migration-accumulation of hydrocarbons, or age of the petroleum system, is based on stratigraphic and petroleum geochemical studies and on the burial history chart. These two processes are followed by the preservation time, which takes place after the generation-migration-accumulation of hydrocarbons occurs, and is the time when hydrocarbons within that petroleum system are preserved, modified, or destroyed. When the generation-migration-accumulation of the petroleum system extends to the present day, there is no preservation time, and it would be expected that most of the petroleum is preserved and that comparatively little has been biodegraded or destroyed. The last event is the critical moment as determined by the investigator from the **burial history** chart, and it shows the time represented on the map and cross section.

Level of Certainty

A petroleum system can be identified at three levels of certainty: known, **hypothetical**, or **speculative**. The level of certainty indicates the confidence for which a particular pod of active source rock has generated the hydrocarbons in an accumulation. In a known petroleum system, a good geochemical match exists between the active source rock and the oil or gas accumulations. In a hypothetical petroleum system, geochemical information identifies a source rock, but no **geochemical match** exists between the source rock and the petroleum accumulation. In a speculative petroleum system, the existence of either a source rock or petroleum is postulated entirely on the basis of geologic or geophysical evidence.

Evolution of a Petroleum System

The time of hydrocarbon generation for a petroleum system can span considerable time and cover a large area. The time span over which petroleum generation occurs can be determined for a series of locations to show how the petroleum system evolves in time and space. At given time increments within this time span, maps and cross sections can be drawn to show the kinematic evolution of the petroleum system. Knowing the age of various horizons within the overburden rock is the key to determining when and where a source rock first starts generating petroleum and when and where it finishes generating petroleum. For example, for a petroleum system whose overburden rock has been deposited over a broad area (such as a prograding deltaic sequence), the time span over which petroleum generation-migration-accumulation occurs is quite large. If this deltaic sequence, which is the overburden rock, has prograded over a 50 Ma period from west to east, then the underlying source rock in this petroleum system will generate petroleum first on the west and last on the east. The geologist knows that it is not always practical to show from start to finish the kinematic development of this petroleum system in 50 Ma increments, as it would require up to 11 maps and cross sections. However, one map and cross section can be drawn to represent the time when the west end of the cross section shows the source rock at maximum burial depth, and an other map and cross section can be drawn to represent the time when the east end of the cross section shows the source rock at maximum burial depth. If more detail is required to better understand how the system evolved, then additional maps and cross sections can be drawn.

New Words and Phrases

1. a hydrocarbon accumulation 一个油气藏(田)
2. biogenic gas 生物气
3. conventional reservoirs 常规油气藏，常规储集岩
4. gas hydrate 天然气水合物

5. tight reservoir 致密气藏
6. condensate *n.* 凝析油
7. asphalt *n.* 沥青
8. overburden rock 上覆岩层
9. burial history 埋藏史
10. seep *n.* & *v.* 油气苗，渗漏
11. hypothetical *adj.* 可能的，假定的
12. speculative *adj.* 推测的，推理的
13. geochemical match 地球化学可比性

Lesson 30 ▸ Play and Prospect

Lesson 30

The distribution of hydrocarbons in the Earth's crust follows a **lognormal distribution** typical of many other natural resources. Such a distribution implies that hydrocarbons are concentrated in relatively few basins, and that exploration is not an **equal-chance game**. In our assessment process, we evaluate four different concepts of exploration as a function of the degree of knowledge about the specific project: sedimentary basin, petroleum system, play, and prospect (Figure 30.1).

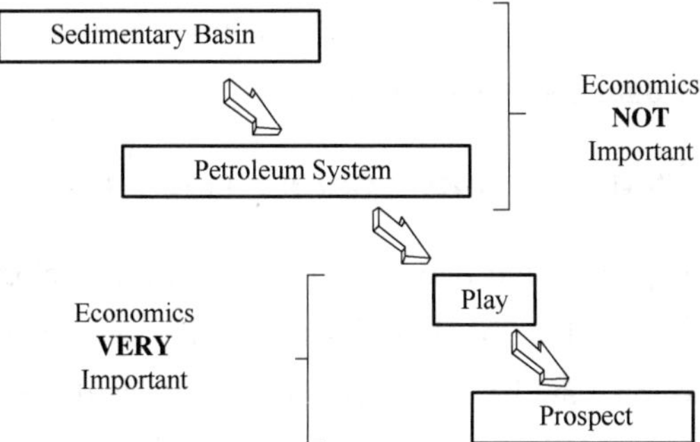

Figure 30.1　Four levels of petroleum investigation

Sedimentary Basin

Is there a volume of sedimentary rocks capable of containing potential ingredients of a working "hydrocarbon machine": source, reservoir, trap and seal, and proper timing and migration? This assessment is a **screening device** only, and does not include economic considerations.

Petroleum System

The petroleum system framework is defined as a volume of sedimentary rocks containing hydrocarbons and charged by a single source rock. The definition requires manifestations of hydrocarbons (seeps, shows, or a producing well) and is applicable in many **frontier basins** only by analogy. Recognition of an active petroleum system also serves only as a screening device because it carries no volumetric (and therefore, no economic) value.

Play

In our definition, the **play** is the elemental part of a petroleum system, and is recognized as having one or more accumulations of hydrocarbons identified by a common geological character of reservoir, trap, and seal; migration; preservation; a common engineering character of location, environment, and fluid and flow properties; or a combination of these. Individual plays, therefore, have unique geological and engineering features, and can be used as a basis for economic characterization.

Prospect

Prospect represents an individual, potential accumulation. Each prospect is perceived as belonging to an individual play, characterized by risk components and a probabilistic range distribution of potential hydrocarbon volumes within its trap confines. In frontier areas, geological analogs provide the best models for assessing the capability of the evaluated basin to yield commercial accumulations of hydrocarbons. In more mature areas, the presence of a petroleum system has been proven, and the assessment focuses on play types. Regardless of the maturity of exploration or the amount of existing production, however, each prospect requires a detailed review of the individual risk components.

The Play as an Economic and Commercial Unit

To be of value to the explorer, the play should have both a geologic as well as an economic implication. Ideally, it should thus be based on families of existing fields and potentially commercial discoveries, which can be used as analogs to guide the economic evaluation of undrilled prospects that seem to belong to the same play. This is probably the most widespread use of the concept—not only is analysis of the chance of success in individual prospects facilitated, but whole play-based strategies can be established and monitored. Many oil or gas indications or occurrences of petroleum, however, are currently noncommercial and do not have an economic status. Although some of these should probably not, strictly speaking, be considered as plays, many of them will have the potential to become commercial in the future, for instance, through developments in technology, so they should not be ignored. Such accumulations commonly point to the existence of a working petroleum system and a combination of factors which, under more favorable circumstances, may result in a commercial accumulation. A separate category of proven play should be introduced for these currently noncommercial plays.

Categories of plays are commonly defined with reference to their status in the basin, province, or petroleum system in which they occur. The following categories may be recognized:

(1) **Speculative plays**, where the play may not yet have been identified in a particular area but, either conceptually or by analogy with similar situations in other similar basins, it is considered that it may be present. This category will typically be associated with frontier provinces where data are sparse, will depend on the identification of appropriate analogs, and will carry a correspondingly low chance of success. The tectonostratigraphic stage in basin evolution provides a valuable context here.

(2) **Unproven plays**, where a play may have been identified in an area and the concept

shown to work, but as yet it has not yet been shown by drilling to be commercial. This category may also be associated either with frontier areas or with an underexplored play in a producing area. This play will have a greater chance of success, and successful analogs from elsewhere will be used to estimate its possible value.

(3) **Emerging plays**, which have been shown to be commercial, include productive fields and, possibly, potentially commercial prospects. Although the play is proven, each prospect will carry several risks specific to the prospect. Identification of these will rely on detailed knowledge of productive fields.

(4) **Mature plays**, proven plays that have had a long period of production, representing a fairway or trend that is in an advanced stage of exploration and development. The number of undrilled prospects, or complementary play opportunities will typically be small.

The first two represent exploration plays, the last two are proven plays. In recent years, the play concept has been increasingly used to categorize investment opportunities other than purely geologic ones. The term is sometimes applied to opportunities represented by new technologies such as secondary or tertiary recovery in existing fields, whereas in other cases, the commercial aspect is dominant. Thus, for instance, Leonard and Ozkaynak classified elements in Shell's high-level strategic themes, Deepwater, Exploration and Production (E&P) Gas, E&P Oil Focus, and Former Soviet Union (FSU)/Russia as plays. Each play then comprised several project elements in the Shell portfolio of global exploration and production opportunities. Clearly, the use here is fundamentally different to that discussed above.

New Words and Phrases

1. play *n.* 成藏组合
2. prospect *n.* 远景圈闭
3. lognormal distribution 正态分布
4. equal-chance game 等概率游戏
5. screening device 筛选方法,筛选手段
6. frontier basin 边际盆地
7. speculative play 推测的成藏组合
8. unproven play 未探明成藏组合
9. emerging play 新兴成藏组合
10. mature play 成熟成藏组合

Lesson 31 ▸ Three Steps in Seismic Exploration

Data Acquisition

Seismic surveys are carried out on land and at sea in different ways. On land the energy source may be provided by detonating explosives buried in **shot holes**, by dropping a heavy weight off the back of a lorry (the thumper technique is actually a rather sophisticated procedure), or by vibrating a metal plate on the ground. The returning acoustic waves are recorded on **geophones arranged in groups**. The signals are transmitted from the geophones along cables to the recording truck. Equipment in this truck controls the firing of the energy source and records the incoming signals from the geophones on magnetic tapes.

The shot points and the receiving geophones may be arranged in many ways. Many groups of geophones are commonly on line with shot points at the end or in the middle of the geophone spread. Today, **common depth point**, or CDP, coverage is widely used. In this method the shot points are gradually moved along a line of geophones. In this way up to 48 signals may be reflected at different angles for a common depth point.

For deep exploration the air gun is a widely used energy source. In this method a bubble of compressed air is discharged into the sea; usually a number of energy pulses are triggered simultaneously from several air guns. The air guns can emit energy sufficient to generate signals at between 5s and 6s two-way travel time. Depending on interval velocities, these signals may penetrate to over 5km.

The reflected signals are recorded by hydrophones on a cable towed behind the ship. The cable runs several meters below sea level and may be between 2km and 5km length. As with land surveys the CDP method is employed and up to 48 recorders may be used. The reflected signals are transmitted electronically from groups by **hydrophones** along the cable to the recording unit on the survey ship.

Other visual equipment on the ship includes a **fathometer** and position-fixing devices. The accurate location of shot points at sea is obviously far more difficult than it is on land. It is generally done either by radio positioning or by getting fixes on two or more navigation beacon transmitters from the shore, backed up by a satellite navigation system.

Data Processing

Once seismic data has been acquired, it must be processed into a format suitable for geological interpretation. This process involves the statistical manipulation of vast numbers of data using mathematical techniques far beyond the comprehension of geologists. Seismic data processors include mathematicians, physicists, electronic engineers, and computer programmers.

Consider the signals of three receivers arranged in a straight line from a shot point, the time taken for a wave to be reflected from a particular layer will appear as a wiggle on the receiver trace. The arrival times will increase with the distance of receivers from the shot point.

With common depth point shooting, the signals are combined, or **stacked**. A signal seismic **reflecting horizon** can be traces across it. **Wave traces** are displayed in several ways. They may be shown as a straightforward graph or wiggle trace. This type of display is the basic for the conventional seismic section used today.

Four main steps are involved in the processing of raw seismic data before the production of the final seismic section:

(1) Conversion of field magnetic tape data into a state suitable for processing.

(2) Analysis of data to select optimum processing parameters.

(3) Processing to remove multiple reflectors and enhance **primary reflectors**.

(4) Conversion of data from digital to analogue form, and printout on graphic display.

One particular important aspect of processing is **wave migration**. When beds dip steeply, the wave returns from the reflector from a point not immediately beneath the surface location midway between the shotpoint and each individual geophone but from a point up-dip from this position. During processing, the data must be migrated to correct this effect. This migration causes several important modifications in the resultant seismic section. Anticlines become sharper, synclines gentler, and faults more conspicuous.

Date Interpretation

Geologists looking at seismic lines and maps inevitably tend to see them as representations of rock and forget that they do, in fact, represent time. Three main groups of pitfalls in seismic interpretation are to be avoided:

(1) Pitfalls due to processing;

(2) Pitfalls due to local velocity anomalies;

(3) Pitfalls due to rapid changes in geometry.

Pitfalls due to processing are the most difficult for geologists to avoid. One of the most common processing pitfalls is multiple reflections—a series of parallel reflections caused by the reverberation between two reflectors. The most prevalent variety is the seabed multiple, which, as its name states, is caused by reverberation between the sea surface and the sea floor. Harder to detect are multiple caused by events within the sediments themselves. These events generally occur where there is a formation with high reflection coefficients above and below. Multiples can be removed by deconvolution and filtering during processing.

Rapid local variation in formation velocity causes many pitfalls. Two of the best-known examples of these variations are produced by salt domes and reefs. In both cases salt and limestone may have faster travel times than the adjacent sediments. Thus the **pre-salt** or **pre-**

reef reflector may appear on the seismic section as an apparent anticline, when it is, in fact, a **velocity pull-up**. A number of salt domes were drilled in the quest for such fictitious anticlines until the method of undershooting them was developed as an effective aid to mapping pre-salt structure. Less serious a pitfall is the sub-reef "anticline", since the reef itself is a valid petroleum prospect and the well will probably be scheduled to drill to pre-reef rocks anyway.

On a regional scale, units may sometimes apparently thin with increasing depth from shelf to basin floor. The formations may, in fact, have a constant thickness, but their velocities increase with increasing compaction. Thus the time taken for seismic waves to cross each interval decreases basinward.

The third group of pitfalls in seismic interpretation occur because of the departure of rock geometry from a simple layered model. This pitfall may cause reflectors to dip steeply, as in tight folds and diapirs, or even to terminate. The former can give rise to distorted reflections, such as the **bow tie effect** produced by synclines. Reflector terminations cause diffractions, which cross-cut the reflectors on the seismic line.

The time for the reflectors for each shot point may then be tabulated. **Contour maps** may then be drawn for each horizon. Note that these are time maps, not depth maps. They show isochrons (lines of equal two-way travel time), not structure contours.

New Words and Phrases

1. shot hole　炮眼,炮井
2. geophones arranged in groups　成组排列的检波器
3. common depth point　共深点
4. hydrophone　*n.* 水中地震检波器
5. fathometer　*n.* 回音测深仪
6. stacked　*adj.* 叠加的
7. reflecting horizon　反射层
8. wave trace　波形
9. primary reflector　一次反射层
10. wave migration　波偏移
11. pre-salt　*n.* 盐下
12. pre-reef　*n.* 礁下
13. velocity pull-up　速度上拉
14. bow tie effect　蝴蝶结效应
15. contour map　等值线图

Lesson 32 ▸ Well Logs

Petrophysical logging, wireline logging, and borehole logging are used as synonyms to describe the same thing. In the same way, the word log is used as a noun, verb, or adjective; here we will limit the term to the act of either recording depth-related information over time (verb) or to the actual recorded hard copy (or digital display) of the depth-based measurement! Logs are used to conduct subsurface correlations, to assist with structure and isopach mapping, help define physical rock characteristics (lithology, porosity, pore geometry, and permeability), and identify production zone thickness and determine the type of formation fluids. There are two basic types of wireline logging tool. Those that are designed to work in an uncased hole and those that are able to work within a cased hole. **Open hole** logs are recorded in the uncased section of the well, usually at some intermediary depth or total depth (TD), prior to running casing. This is because some logging tools do not work well in cased holes. For example, microresistivity logging devices cannot work through casing, unlike acoustic (or sonic) tools. A useful basis for subdividing open hole wireline logging tools is on the basis of type or end-use, which will be used here.

Spontaneous Potential (SP) Log

The SP log is used to identify impermeable (e.g., shale) and permeable zones (e.g., sand), and help determine R_w (resistivity of formation water). This tool can only be used in the open hole containing a conductive drilling fluid (i.e., not **oil-based mud**). The SP log is a record of the DC voltage difference between the naturally occurring potential of a moveable electrode in the well bore and the potential of a fixed electrode located at the surface. The SP is influenced by bed thickness, bed resistivity, invasion, borehole diameter, and shale content. The SP device has a **vertical resolution** of 5.5m to 9.3m.

Gamma Ray Log

The gamma ray curve is also placed in track 1. The scale, is given in API units, typically ranges from 100 to 150. Tracks 2 and 3 often contain porosity logs or resistivity logs. The gamma ray log is affected by borehole dimension, but because the gamma ray tool detects the presence of natural radioactivity it can be used in cased hole. The resolution of the gamma ray device is about 0.6m with a radius of investigation half that value, the gamma ray tool is typically combined with other devices.

Sonic Log

This is an open hole, borehole compensated, porosity log that measures the interval transit time (Δt) of a compressional sound wave through the formation. The interval transit time (Δt) is the reciprocal of matrix velocity and is therefore expressed in microseconds per unit distance. The acoustic "clicks" emitted from the transmitter travel faster through formation than through the drilling fluid and are picked up by the receiver located on the same tool. Because interval transit time is dependent upon rock type and rock density (i. e., mineral density and cement), and also upon the fluid within the pores, it can, therefore, be related to formation porosity. Sonic logs (like all logs) are best used in conjunction with other logs (e. g., density, SP, resistivity). Sonic logs also work best in consolidated and compact formations. There are two types of sonic devices: the compensated compression wave sonic device and the full waveform sonic (FWS) device. The FWS device contains an array of receivers designed to detect shear velocities, this log is used by log analysts to determine the mechanical properties of the lithologies of interest. Sonic devices have a vertical resolution of about 0.5m with a radius of investigation of about 12cm.

Formation Density Log

The formation density tool is another pad-type device that measures the electron density of a formation. This tool consists of source (**cobalt**-60 or **cesium**-137) that emits medium-energy gamma rays into the formation. The gamma rays collide with electrons in the formation and lose energy and/or are scattered; redirected as a photon of reduced energy, scattered energy is detected by the device. This device is used to determine formation porosities when lithology is known, determine lithology in conjunction with other logs, detect gas-bearing zones, and identify uncommon minerals in evaporite beds.

Nuclear Magnetic Resonance Log

The **nuclear magnetic resonance** (NMR) device is relatively new and provides a lithology-independent measure of effective porosity, total porosity, and differentiates between irreducible and moveable (i. e., free water).

Formation fluids contain differing amounts of hydrogen that varies with composition (e. g., oil versus water) and physical state (i. e., gas and fluid). All hydrogen nuclei spin, although in the natural state the spin orientation is random and the net magnetization is zero. However, when a strong external magnetic field (from the NMR device) passes through the formation, the protons are aligned or polarized. The state of polarization increases exponentially in time. An antenna in the NMR device emits a pulsating radio frequency that induces a magnetic field causing the spin-axes to tip away from the original (i. e., polarized) alignment. When the antenna is turned off and the magnetic pulse is removed, the resonance relaxes and the signal decays, producing a signal called the spin echo. Repeated pulses create a spin echo train which is used to interpret fluid and formation properties. The polarization time is known as T_p and the time constant used to characterize the magnetization buildup is known as T_1, whereas T_2 represents the transverse decay time, commonly known as relaxation time.

For a given fluid type, T_2 has been shown to be proportional to pore size, in which micropores are associated with the fastest relation times compared to free fluids and fluids within larger pores; in this way NMR logs can be used to differentiate between irreducible and free-water.

Reservoir fluids possess different polarization times and relaxation times. In this way NMR logs can be used to characterize and differentiate between formation fluids (e. g. , oil/water), and for a given formation fluid (i. e. , water) determine the presence of clay-bound, capillary-bound, and free water. More recently, NMR logs have been used to provide estimates of pore-size distribution, grain size (**clastic reservoir**) and estimates of permeability.

New Words and Phrases

1. formation fluids 地层流体
2. open hole 裸眼井
3. oil-based mud 油基钻井液
4. vertical resolution 垂向分辨率
5. nuclear magnetic resonance 核磁共振
6. clastic reservoir 碎屑岩储层

Lesson 33 ▸ Subsurface Maps

Successful petroleum exploration involves the integration of the wireline logs and geophysical surveys with geological data and concepts. This section is written on the assumption that the reader is familiar with the basic principles of stratigraphy, correlation, and contour mapping. Many of the methods to be discussed are not, of course, peculiar to petroleum exploration, but have wider geological applications. Essentially, the two methods of representing geological data are cross-sections and maps. These methods are discussed in the following sections.

Geological Cross-sections

Vertical **cross-sections** are extremely important in presenting geological data. This account begins with small-scale sections and well correlations and proceeds to regional sections.

The starting point for detailed cross-section construction, as for example within a field, is consideration of well correlation. When a well has been drilled and logged, a composite log is prepared. This log correlates the geological data gathered from the **well cuttings** with that of the wireline logs. The formation tops then have to be picked, which is not always an easy task. The geologist, the paleontologist, and the geophysicist may all pick the top Cretaceous, for example, at a different depth. The geologist may pick the top at, say, the first sand bed; the paleontologist, at the first record of a particular microfossil (which may not have thrived in the environment of the sediments in question); and the geophysicist, at a velocity break in the shales, which is a prominent reflecting horizon.

When the formation tops have been selected, correlation with adjacent wells can be made. This correlation is an art in which basic principles are combined with experience. First, the most useful geophysical logs to use must be determined. In the early days of well correlation the **IES array** was the most widely used for composite logs. Now the gamma-sonic is often more popular, especially where the SP curve is ineffective. The gamma, sonic, and resistivity curves all tend to share the high amplitudes necessary for effective correlation. As a general guide, coal beds and thin limestones often make useful markers: they are both thin, but also give dramatic kicks on resistivity and porosity logs.

Sometimes when correlating wells, significant intervals of section may be missing. This phenomenon may be caused by **depositional thinning**, erosion, or normal faulting. Examination of the appropriate seismic lines will generally reveal which of these three possibilities is the most likely. Repetition of sections may sometimes be noted. Repetition may be caused by reverse faulting, but this possibility should only be considered in regions known to have been subjected to compressional tectonics. An alternative explanation is that sedimentation is cyclic, causing the repetition of log motifs.

Constructing Cross-sections

Once wells have had their formation tops picked and they have been correlated, they may be hung on a cross-section. This procedure is done with a datum, which may be sea level, a fluid contact, or a particular geological **marker horizon**. When sea level is used, the elevation of the log must be adjusted. Log depths are generally measured below the rotary table or **kelly bushing**. The elevation of whichever one of these was used is given on the log heading. The elevation above sea level is subtracted from each formation top to find its altitude or depth. This procedure is relatively simple for vertical wells, but less so for deviated wells, such as those drilled from a marine production platform. For these types of wells the true vertical depth (TVD) must be determined, which requires a detailed and accurate knowledge of the path of the borehole.

When a cross-section is drawn to a horizontal datum, be it sea level or an oil-water contact, the result is a structural cross-section. Alternatively, a cross-section can be constructed using a geological horizon as a datum. The purpose of the cross-section should be considered before the datum is selected.

Cross-sections of fields are generally based on well control. For regional studies a combination of seismic and well data is used. A development of the single cross-section is a series drawn using a sequence of different data horizons. Where these data horizons are selected to span a number of markers up to the present day, they can be used to document the evolution of a basin or an individual structural feature.

Many different types of subsurface geological maps are used in oil exploration. The simplest, and probably most important, subsurface map is the **structure contour map**. This map shows the configuration of a particular horizon with respect to a particular datum, generally sea level. Structure contour maps may be regional or local, indicating the morphology of basins and traps, respectively, and delineate traps and are essential for reserve calculations.

Next in importance are isopach maps, which record the thickness of formations. As with structural maps they may be regional or local and tend to be most reliable when seismic and well data are integrated. Care must be taken when interpreting isopach maps. There is a natural tendency to think that thickness increases with basin subsidence, which is not necessarily so. Both sand and carbonate sedimentation rates are often at a maximum a short way into a basin. Beds thin toward the shore because of erosion and basinward because of nondeposition or slow sedimentation rates.

Net sand maps may often be combined with paleogeographic maps. Paleogeographic maps should be based on seismic and well data from which the depositional environment has been interpreted. This data may then be used to delineate the paleogeography of the area during the deposition of the sediments studied. Such maps can be used to predict the extent and quality of source rocks and reservoirs across a basin.

Another useful type of map is the subcrop, or **pre-unconformity map**, which can be constructed from seismic and well data. Regional **sub-crop maps** are useful for showing where reservoirs are overlain by source rocks and vice versa. On a local scale subcrop maps may be combined with isopach maps to delineate truncated reservoirs. Subcrop maps also give some indication of the structural deformation imposed on the underlying strata.

New Words and Phrases

1. cross-section *n.* 剖面图
2. well cutting 钻井岩屑
3. IES (induction electrical survey) array 感应测井曲线
4. marker horizon 标志层
5. depositional thinning 沉积减薄
6. kelly bushing 方钻杆补心
7. structure contour map 构造等高线图
8. sub-crop map 地下地质图
9. pre-unconformity map 不整合面下地质图

Lesson 34 ▸ Evaluation of Oil & Gas Resources

The authors summarize a **geology-based assessment** of undiscovered conventional oil and gas resources of priority geologic provinces of the world, completed between 2009 and 2011 as part of the U. S. Geological Survey (**USGS**) World Petroleum Resources Project. One hundred seventy-one geologic provinces were assessed in this study (exclusive of provinces of the United States), which represent a complete reassessment of the world since the last report was published in 2000. The present report includes the recent oil and gas assessment of geologic provinces north of the Arctic Circle. However, not all potential oil-and gas-bearing provinces of the world were assessed in the present study.

The methodology for the assessment included a complete geologic framework description for each province based mainly on published literatures, and the definition of petroleum systems and assessment units (AU) within these systems. In this study, 313 AU were defined and assessed for undiscovered oil and gas accumulations. Exploration and discovery history was a critical part of the methodology to determine sizes and numbers of undiscovered accumulations. In those AUs with few or no discoveries, geologic and production analogs were used as a partial guide to estimate sizes and numbers of undiscovered oil and gas accumulations, using a database developed by the USGS following the 2000 assessment. Each AU was assessed for undiscovered oil and **nonassociated gas accumulations**, and co-product ratios were used to calculate the volumes of associated gas (gas in oil fields) and volumes of natural gas liquids. This assessment is for conventional oil and gas resources only; unconventional resource assessments (heavy oil, tar sands, shale gas, shale oil, tight gas, coalbed gas) for priority areas of the world are being completed in an ongoing but separate USGS study.

The USGS assessed undiscovered conventional oil and gas resources in 313 AU within 171 geologic provinces. In this report the results are presented by geographic region, which correspond to the eight regions used by the U. S. Geological Survey World Energy Assessment Team (2000). For undiscovered, technically recoverable resources, the mean totals for the world are as follows: (1) 565,298 million barrels of oil (MMBO); (2) 5,605,626 billion cubic feet of gas (BCFG); and (3) 166, 668 million barrels of natural gas liquids (MMBNGL). The assessment results indicate that about 75 percent of the undiscovered conventional oil of the world is in four regions: South America and Caribbean, **sub-Saharan Africa**, Middle East and North Africa, and the Arctic provinces portion of North America. Significant undiscovered conventional gas resources remain in all of the world's regions.

The potential additions to reserves from **reserve growth** (Figure 34. 1) are nearly as large as the estimated **undiscovered resource volumes**. These estimates imply that 75 percent of the world's **grown conventional oil endowment** and 66 percent of the world's grown conventional gas endowment have already been discovered in the areas assessed (exclusive of the U. S.). Additionally, for these areas, 20 percent of the world's grown conventional oil endowment and 7 percent of the world's grown conventional gas endowment had been produced as of the end of 1995.

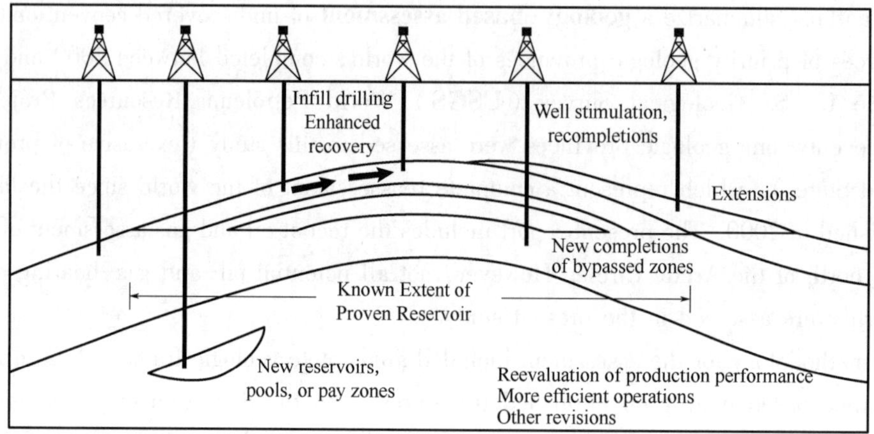

Figure 34.1 Diagram showing reserve growth defined as increases in successive estimates of recoverable quantities of crude oil, natural gas, and natural gas liquids in discovered fields. Reserve growth can be grouped into three activities: (1) delineation of **additional in-place petroleum volumes**, which increases the degree of geologic assurance (**infill drilling**; new reservoirs, pools, or pay zones; extensions); (2) improved recovery efficiency, which increases the degree of technological feasibility (**enhanced recovery**, well stimulation, recompletions, new completions of **bypassed zones**); and (3) revisions resulting from recalculation of viable reserves in dynamically changing economic, operating, and regulatory/political conditions, which increases the degree of economical feasibility (reevaluation of production performance; more efficient operations) (Klett, 2005).

This assessment is based on extensive geologic studies as opposed to statistical analysis. A team of more than 40 geoscientists and additional supporting staff conducted the study over a five-year period from 1995 to 2000. The petroleum assessed occurs in fields exceeding a stated minimum size, which varies between 1 million barrels and 20 million barrels of **oil equivalent** in different areas, and in accumulation categories judged to be viable in a 30-year forecast span.

The critical geologic controls on petroleum distribution are encompassed by the Total Petroleum System (TPS) and were studied using this approach. Assessment Units (AU), within the TPS, were the basic units for assessment. The assessed areas were those judged to be significant on a world scale in terms of known petroleum volumes, geologic potential for new petroleum discoveries, and political or societal importance. In the course of our geologic analyses, 24 AU were identified as containing **continuous (nonconventional) resources**.

For each AU, allocations of undiscovered resources were made to the countries, geologic provinces, regions, and offshore areas (if any) involved. From these allocated portions, aggregations of estimates were made for higher levels such as to countries, geologic provinces, and groups of countries including the Organization of the Petroleum Exporting Countries (OPEC) and the Organization for Economic Co-operation and Development (OECD).

Areas assessed in the World Petroleum Assessment 2000 that contain the greatest volumes of undiscovered conventional oil include the Middle East, northeast **Greenland Shelf**, the West Siberian and Caspian areas of the Former Soviet Union, and the Niger and Congo delta areas of Africa. Significant new undiscovered oil resource potential was identified in a number of areas with no significant production history, such as northeast Greenland and **offshore Suriname**.

Areas that contain the greatest volumes of undiscovered conventional gas include the West Siberia Basin, Barents and Kara Seas shelves of the Former Soviet Union, the Middle East, and offshore Norwegian Sea. A number of areas were identified that may contain significant additional undiscovered gas resources where large discoveries have been made but remain undeveloped. Examples include East Siberia and the Northwest Shelf of Australia.

Results of USGS World Petroleum Assessment 2000 offer opportunities for many studies beyond the initial analyses in this report. The material generated by this effort can serve as a foundation for additional geologic, economic, geopolitical, and environmental studies.

New Words and Phrases

1. geology-based assessment 基于地质的评价
2. USGS *abbr.* 美国地质调查局
3. nonassociated gas accumulations 非伴生气藏
4. sub-Saharan Africa 撒哈拉以南非洲
5. reserve growth 储量增长，扩展储量
6. undiscovered resource volume 未发现资源量
7. grown conventional oil endowment 总可采石油资源量
8. additional in-place petroleum volumes 额外的油气地质储量
9. infill drilling 加密钻井
10. enhanced recovery 提高采收率
11. bypassed zone 未波及层段
12. oil equivalent 油当量
13. continuous (nonconventional) resource 连续型(非常规)资源
14. Greenland Shelf 格陵兰大陆架
15. offshore Suriname 苏里南海域

Lesson 35 ▸ Petroleum Resources Management System

The estimation of petroleum resource quantities involves the interpretation of volumes and values that have an inherent degree of uncertainty. These quantities are associated with development projects at various stages of design and implementation. Use of a consistent classification system enhances comparisons between projects, groups of projects, and total company portfolios according to forecast **production** profiles and recoveries. Such a system must consider both technical and commercial factors that impact the project's economic feasibility, its productive life, and its related cash flows.

Petroleum is defined as a naturally occurring mixture consisting of hydrocarbons in the gaseous, liquid, or solid phase. Petroleum may also contain non-hydrocarbons, common examples of which are carbon dioxide, nitrogen, **hydrogen sulfide** and sulfur. In rare cases, non-hydrocarbon content could be greater than 50%.

The term "resources" as used herein is intended to encompass all quantities of petroleum naturally occurring on or within the Earth's crust, discovered and undiscovered (recoverable and unrecoverable), plus those quantities already produced. Further, it includes all types of petroleum whether currently considered "conventional" or "unconventional."

Figure 35.1 is a graphical representation of the SPE (Society of Petroleum Engineers)/WPC (World Petroleum Council)/AAPG (American Association of Petroleum Geologists)/SPEE (Society of Petroleum Evaluation Engineers) resources classification system. The system defines the major recoverable resources classes: **production**, **reserves**, **contingent resources**, and **prospective resources**, as well as **unrecoverable petroleum**.

The "Range of Uncertainty" reflects a range of estimated quantities potentially recoverable from an accumulation by a project, while the vertical axis represents the chance of commerciality, that is, the chance that the project will be developed and reach commercial producing status. The following definitions apply to the major subdivisions within the resources classification.

Total petroleum initially-in-place is that quantity of petroleum that is estimated to exist originally in naturally occurring accumulations. It includes that quantity of petroleum that is estimated, as of a given date, to be contained in known accumulations prior to production plus those estimated quantities in accumulations yet to be discovered (equivalent to "total resources").

Discovered petroleum initially-in-place is that quantity of petroleum that is estimated, as of a given date, to be contained in known accumulations prior to production.

Production is the cumulative quantity of petroleum that has been recovered at a given date. While all recoverable resources are estimated and production is measured in terms of the sales

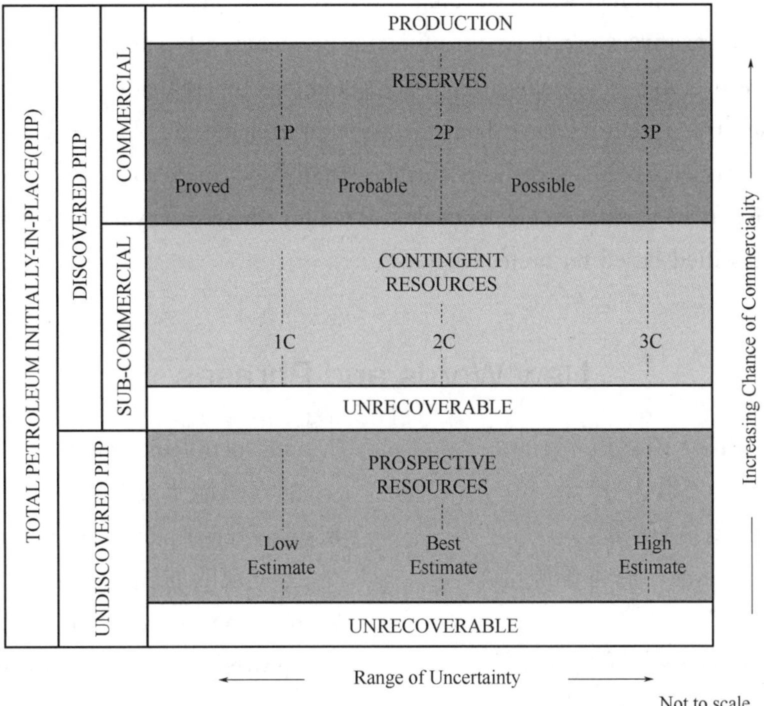

Figure 35. 1 Resources classification framework

product specifications, raw production (sales plus non-sales) quantities are also measured and required to support engineering analyses based on **reservoir voidage**.

Reserves are those quantities of petroleum anticipated to be commercially recoverable by application of development projects to known accumulations from a given date forward under defined conditions. Reserves must further satisfy four criteria: they must be discovered, recoverable, commercial, and remaining (as of the evaluation date) based on the development project(s) applied. Reserves are further categorized in accordance with the level of certainty associated with the estimates and may be sub-classified based on project maturity and/or characterized by development and production status.

Contingent resources are those quantities of petroleum estimated, as of a given date, to be potentially recoverable from known accumulations, but the applied project(s) are not yet considered mature enough for commercial development due to one or more contingencies. Contingent resources may include, for example, projects for which there are currently no viable markets, or where commercial recovery is dependent on technology under development, or where evaluation of the accumulation is insufficient to clearly assess commerciality. Contingent resources are further categorized in accordance with the level of certainty associated with the estimates and may be subclassified based on project maturity and/or characterized by their economic status.

Undiscovered petroleum initially-in-place is that quantity of petroleum estimated, as of a given date, to be contained within accumulations yet to be discovered.

Prospective resources are those quantities of petroleum estimated, as of a given date, to be potentially recoverable from undiscovered accumulations by application of future development projects. Prospective resources have both an associated chance of discovery and a chance of development. Prospective resources are further subdivided in accordance with the level of certainty associated with recoverable estimates assuming their discovery and development and may be sub-classified based on project maturity.

New Words and Phrases

1. hydrogen sulfide 硫化氢
2. production *n.*（累计）产量
3. reserve *n.*（剩余）储量
4. contingent resource 后备资源量，表外储量
5. prospective resource 远景资源量
6. unrecoverable petroleum 不可采油气
7. total petroleum initially-in-place 油气总地质资源量
8. discovered petroleum initially-in-place 已发现的油气地质储量
9. reservoir voidage 油层孔隙空间体积
10. proved reserve 探明储量，证实储量
11. probable reserve 控制储量
12. possible reserve 预测储量

Module 4

Development of Oil-Gas Fields

04

Lesson 36 ▸ Development Geology

Development geology is one of the major branches of petroleum geology, paralleling with petroleum exploration geology. The main objective and content of development geology are to describe the development geology characteristics of reservoir accurately.

After an exploration group discovers a field, the geologic responsibilities for the field are normally turned over to a **development geologist**. The primary function of the development geologist is to develop the field as economically and efficiently as possible. This requires input from **geophysicists**, **exploration geologists**, economists, **reservoir engineers**, drilling engineers, petrophysicists (well-log analysts), and people from other engineering disciplines.

The development geologist works in a subsurface team, a team that manages production for a field and looks for ways of getting more hydrocarbons out of it. He or she has a specific role. The development geologist is responsible for understanding the geological framework of the reservoir and creating a representation of it, typically using **computer software**. The object of this model is to help understand how the geology both influences fluid flow within a producing reservoir and creates dead ends that could potentially trap hydrocarbons. The bigger dead-end pockets may be worth targeting with new wells. If these wells look profitable, the geologist will then take a leading role in planning them with the drilling engineers.

Development geologists build **three-dimensional (3-D) computer models** of the larger fields to represent the geology (Figure 36. 1). The figure shows the relief on the top surface of a reservoir interval. Also shown are the paths of the wells that intersect the top reservoir surface.

Development geologists assigned to an operated field will find themselves working as part of a **multidisciplinary team**. In a large company, this will include a subsurface manager,

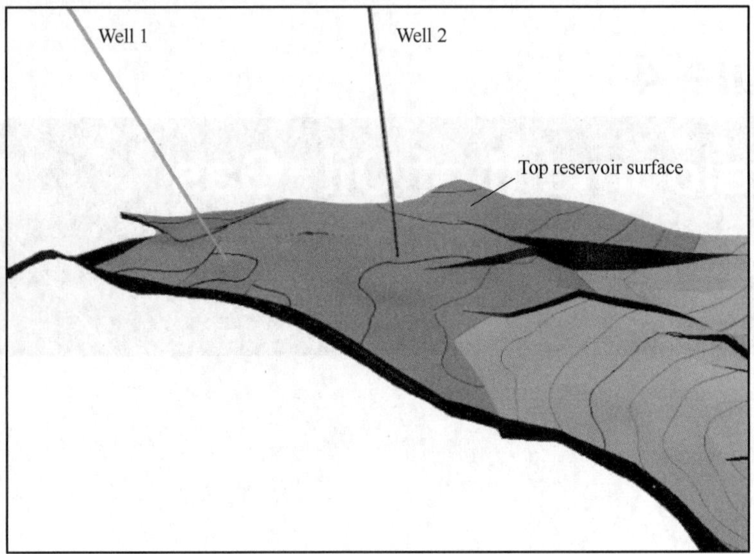

Figure 36. 1 3-D geological model

geologists, geophysicists, petro - physicists, reservoir engineers, production engineers, chemists, and technical assistants. Some teams may also include drilling engineers and economists.

Some oil companies have separate geology and engineering departments, although this rarely works in practice. Short lines of communication should exist within a subsurface team such that an inclusive atmosphere of shared purpose is created. Any problems that arise can then be quickly recognized and solved by common directed action.

The development geologist is in a central position to all of these disciplines and must be able to communicate effectively with all of these people. Teamwork is essential because the staggeringly complex nature of a **subsurface operation** means that the various disciplines have to integrate their specific areas of expertise for the venture to be successful.

Subsurface manager: manages and coordinates the work of everyone in the subsurface team.

Development geologist: responsible for understanding and modeling the **geological framework** of the reservoir. Helps to identify and plan new well locations.

Geophysicist: spends much of his/her time interpreting seismic data to define the **reservoir structure** and **fault distribution**. Where the **seismic data** allow, **depositional environment**, rock, and **fluid properties** can also be characterized.

Petrophysicist: a key task is to analyze **wireline logs** to quantify the rock and fluid properties of the reservoir at the well scale.

Technical assistant: provides technical support to the team. This includes data management, data preparation, and **computer mapping**.

Reservoir engineer: predicts how much oil and gas a field is likely to produce, and may use a computer simulation of reservoir performance to analyze how the field will behave as well as taking a lead in reservoir management activities.

Production engineer: responsible for optimizing all the mechanical aspects of hydrocarbon production from the wellbore to the surface facilities.

Production chemist: analyzes and treats problems related to scale formation, **metal corrosion**, drilling fluids, wax formation, and **solids precipitation** between the reservoir and the surface facilities.

Drilling engineer (well engineer): plans the mechanical aspects of any well operations including drilling new wells.

Economist: costs and evaluates any economic activity relevant to the subsurface.

New words and Phrases

1. development geology 开发地质学
2. development geologist 开发地质学家
3. geophysicist *n.* 地球物理学家
4. exploration geologist 勘探地质学家
5. reservoir engineer 油藏工程师
6. computer software 计算机软件
7. three-dimensional (3-D) model 三维模型
8. multidisciplinary team 多学科团队
9. subsurface operation 地下作业
10. geological framework 地质格架
11. seismic data 地震数据
12. reservoir structure 油藏构造
13. fault distribution 断层分布
14. depositional environment 沉积环境
15. fluid property 流体性质
16. wireline log 测井曲线
17. computer mapping 电脑绘图
18. metal corrosion 金属腐蚀
19. solid precipitation 固体沉淀

Lesson 37

Lesson 37 ▸ Reservoir Drives

For a proper understanding of **reservoir behavior** and predicting future performance, it is necessary to have knowledge of the **driving mechanisms** that control the behavior of fluids within reservoirs. The overall performance of oil reservoirs is largely determined by the nature of the energy, i. e., driving mechanism, available for moving the oil to the **wellbore**. There are basically five driving mechanisms that provide the natural energy necessary for **oil recovery**:

(1) **Rock and liquid expansion drive**;

(2) **Depletion drive**;

(3) **Gas cap drive**;

(4) **Water drive**;

(5) **Gravity drainage drive**.

These driving mechanisms are discussed as follows.

Rock and Liquid Expansion Drive

When an oil reservoir initially exists at a pressure higher than its bubble-point pressure, the reservoir is called an **undersaturated oil reservoir**. At pressures above the bubble-point pressure, crude oil, connate water, and rock are the only materials present. As the reservoir pressure declines, the rock and fluids expand due to their individual compressibilities. The reservoir rock compressibility is the result of two factors:

(1) Expansion of the individual rock grains;

(2) Formation compaction.

Both of the above two factors are the results of a decrease of fluid pressure within the pore spaces, and both tend to reduce the pore volume through the reduction of the porosity.

As the expansion of the fluids and reduction in the pore volume occur with decreasing reservoir pressure, the crude oil and water will be forced out of the pore space to the wellbore. Because liquids and rocks are only slightly compressible, the reservoir will experience a rapid pressure decline. The oil reservoir under this driving mechanism is characterized by a constant gas-oil ratio that is equal to the gas solubility at the bubble point pressure.

This driving mechanism is considered the least efficient driving force and usually results in the recovery of only a small percentage of the total oil in place.

Depletion Drive

In this type of reservoir, the principal source of energy is a result of gas liberation from the crude oil and the subsequent expansion of the solution gas as the reservoir pressure is reduced. As pressure falls below the bubble-point pressure, gas bubbles are liberated within the microscopic pore spaces.

The reservoir pressure declines rapidly and continuously. Gas - oil ratio increases to maximum and then declines. There will be little or no water production with the oil during the entire producing life of the reservoir. Ultimate oil recovery from depletion-drive reservoirs may vary from less than 5% to about 30%.

Gas Cap Drive

Due to the ability of the gas cap to expand, these reservoirs are characterized by a slow decline in the reservoir pressure. The natural energy available to produce the crude oil comes from the following two sources:

(1) Expansion of the gas-cap gas;

(2) Expansion of the solution gas as it is liberated.

The reservoir pressure falls slowly and continuously. There will be absent or negligible water production. The gas-oil ratio rises continuously in up-structure wells. The expected oil recovery ranges from 20% to 40%.

Water Drive

Many reservoirs are bounded on a portion or all of their peripheries by water bearing rocks called **aquifers**. The aquifers may be so large compared to the reservoir they adjoin as to appear infinite for all practical purposes, and they may range down to those so small as to be negligible in their effects on the reservoir performance.

The aquifer itself may be entirely bounded by **impermeable rock** so that the reservoir and aquifer together form a closed (volumetric) unit. On the other hand, the reservoir may be outcropped at one or more places where it may be replenished by surface water.

It is common to speak of **edge water** or **bottom water** in discussing water influx into a reservoir. Bottom water occurs directly beneath the oil and edge water occurs off the flanks of the structure at the edge of the oil. Regardless of the source of water, the water drive is the result of water moving into the pore spaces originally occupied by oil, replacing the oil and displacing it to the producing wells.

The reservoir pressure remains high and surface gas - oil ratio remains low. Water production starts early and increases to appreciable amounts. The ultimate oil recovery normally ranges from 35% to 75% of the original oil in place.

Gravity Drainage Drive

The mechanism of gravity drainage occurs in petroleum reservoirs as a result of differences in densities of the reservoir fluids. The effects of **gravitational forces** can be simply illustrated by placing a quantity of crude oil and a quantity of water in a jar and agitating the contents. After agitation, the jar is placed at rest, and the more dense fluid (normally water) will settle to the bottom of the jar, while the less dense fluid (normally oil) will rest on top of the denser fluid. The fluids have separated as a result of the gravitational forces acting on them.

The fluids in petroleum reservoirs have all been subjected to the forces of gravity, as evidenced by the relative positions of the fluids, i. e., gas on top, oil underlying the gas, and water underlying oil. Due to the long periods of time involved in the petroleum **accumulation-and-migration** process, it is generally assumed that the reservoir fluids are in equilibrium. If the reservoir fluids are in equilibrium, then the gas-oil and oil-water contacts should be essentially horizontal. Although it is difficult to determine precisely the reservoir fluid contacts, best available data indicate that, in most reservoirs, the fluid contacts actually are essentially horizontal.

Gravity segregation of fluids is probably present to some degree in all petroleum reservoirs, but it may contribute substantially to oil production in some reservoirs.

The reservoir shows a rapid pressure decline, low gas-oil ratio from structurally low wells, little or no water production. The ultimate oil recovery normally has exceeded 80% of the initial oil in place.

New Words and Phrases

1. reservoir behavior　储层动态
2. driving mechanism　驱动机制
3. wellbore　*n.* 井筒
4. oil recovery　采油
5. rock and liquid expansion drive　岩石和流体的弹性驱动
6. depletion drive　溶解气驱
7. gas cap drive　气顶气驱动
8. water drive　水驱
9. gravity drainage drive　重力排水驱动
10. undersaturated oil reservoir　未饱和油藏
11. pore space　孔隙空间
12. gas-oil ratio　气油比
13. gas bubble　气泡
14. ultimate oil recovery　油层最终采收率
15. water production　产水
16. aquifer　*n.* 含水层
17. impermeable rock　不渗透岩层
18. edge water　边水
19. bottom water　底水
20. gravitational forces　重力
21. accumulation-and-migration　成藏和运移

Lesson 38 ▸ Enhanced Oil Recovery (EOR)

Lesson 38

Historically, **water flooding** and **gas injection** have been referred to as **secondary recovery techniques** and other, more exotic, techniques have been referred to as **tertiary techniques**. Today, the term **enhanced oil recovery** (EOR) includes both secondary and tertiary recovery techniques.

The worldwide average recovery efficiency for primary oil is on the order of 33 percent. This means that, after primary recovery, two thirds of the oil is left in the ground, principally as **residual oil** in **water wet** reservoirs.

Secondary Recovery Techniques

Water Flooding

Water flooding is used on a routine basis throughout the world to maintain reservoir pressure and to push oil in front of a water front. Injected water is normally taken from the subsurface because surface waters or seawater commonly react with formation waters to cause undesirable **precipitates** or expansion of **clay minerals**. Whatever the origin of the water, its chemistry must be checked carefully against the chemistry of the formation fluids to make certain that the fluids are chemically compatible. It can be quite embarrassing to start injecting water only to find that the formation is now **impermeable** because of precipitates such as $BaSO_4$ or other **insoluble** minerals caused by mixing of incompatible fluids. Such insoluble precipitates can virtually shut off permeability around injection wells.

Configuration of the reservoir, depth, and cost are important factors in determining which type of injection pattern to implement in a water flood. Typically, for thin, steeply dipping reservoirs, an edge water drive is expected. Water injectors are placed near the original water level and producing wells are placed updip. As the oil-water contact moves **updip** and producing wells water out, the production wells are progressively converted to water injectors.

Gas Injection

For gas reservoir associated with oil, it is of course undesirable (illegal in some cases) to blow down the gas cap before the oil is produced. Wells that produce from near the gas-oil contact commonly produce with a high **gas-oil ratio** (**GOR**) as gas is coned downward into the oil **perforations**. In order to maintain reservoir pressure, solution gas is commonly compressed and injected back into the gas cap. Alternatively, the compressed gas may be injected into the annulus, where it passes through gas lift mandrels into the tubing. This injected gas helps lighten the oil column in the tubing, which in turn helps lift the oil to the surface. Gas lift, like pumping, is a **primary oil recovery** technique.

Tertiary Recovery Techniques

Chemical floods

(1) **Polymer flooding**: Polymers are commonly added to injection water to make the water more **viscous**, thus reducing the water's mobility. This tends to plug up the high-permeability zones which will normally improve sweep efficiency. Foam is another way to decrease the mobility of the displacing fluid and create a more uniform sweep efficiency.

(2) **Surfactants** flooding: Oil and water are **immiscible**, meaning that they form **emulsions** in the subsurface that can be very difficult to break. Emulsions are caused by high **surface tension** between the two fluids. Surfactants such as soap or micelles are commonly added to polymer floods to break the emulsions and decrease interfacial tension between oil and water. **Caustic soda** is used to increase the pH of the reservoir, and depending on the type of crude, react with the oil to create surfactants that help move the oil through the pore throats. Caustic flooding is normally applied to relatively acid crudes with high **API gravity** numbers.

Thermal Recovery Processes

Steam flooding consists of two principal types. In relatively small reservoirs or reservoirs that do not have good **lateral permeability**, a steam soak is commonly used. Steam is injected into the reservoir and allowed to soak into the formation for a period of time, commonly a week. The steam heats the oil and reduces its viscosity. The injection well is then produced for a period of time, also commonly a week. This procedure is referred to as a "**huff and puff**" process, as the well or wells are alternately used as injection wells and then as producing wells. In a conventional steam flood, steam is injection through injection wells and production occurs through production wells. Usually, this is a patterned flood such that the entire reservoir is swept. Surfactants are commonly added to help mobilize the oil. Steam floods are normally used at relatively shallow depths because, as pressure increases with depth, higher temperatures are required to keep water in vapor form.

In-situ combustion is a process whereby air or oxygen is injected into the formation where combustion of reservoir oil and gas can occur. The heat generated by the combustion creates a steam bank that drives the oil to producing wells.

Hot water injection is much like a water flood, except that the water is heated to reduce the viscosity of the oil.

Electromagnetic or heating by microwaves has been considered by a number of companies. Microwaves are very efficient at heating surfaces, but they do not penetrate more than a few centimeters past the surface. To date, no one has discovered a method to transmit the energy deep into the formation where it can do some real good in terms of mobilizing oil.

Miscible Hydrocarbon Displacement

Oil and some natural gases, such as ethane, are miscible, meaning that the surface tension between the two phases is very low. Some natural gases tend to process can significantly help move relatively heavy and viscous oils through pore throats. Miscible floods tend to be quite expensive, but they can be attractive if the majority of the injected gas is ultimately recovered.

Carbon dioxide, nitrogen, and flue gases are the most commonly injected gases because they are cheap, they are often readily available as waste products, and they reduce the surface tension of the oil.

Microbial EOR

To date, no large-scale microbial projects have been attempted. There are two principal ideas behind **microbial EOR**. In the first method, microbes plus nutrients are injected into the formation. The microbes decompose the oil to produce detergents, CO_2, and new cells which either mechanically or chemically release oil from the reservoir pores.

In the second method, microbes and nutrients are injected into the reservoir, where they partially degrade the oil. Through this mechanism, the degraded oil and microbes block off areas of highest permeability such that further injection of other fluids causes the zones of lower permeability to be selectively flushed.

New Words and Phrases

1. water flooding 注水
2. gas injection 注气
3. secondary recovery techniques 二次采油技术
4. tertiary techniques 三采技术
5. enhanced oil recovery (EOR) 提高原油采收率
6. residual oil 残油
7. water wet 亲水的
8. precipitate *n.* 沉淀物
9. clay mineral 黏土矿物
10. impermeable *adj.* 不渗透的
11. insoluble *adj.* 不溶解的
12. updip *vi.* & *adj.* 上倾(的)
13. gas-oil ratio (GOR) 气油比
14. perforation *n.* 孔隙
15. primary oil recovery 初次采油
16. chemical flood 化学驱油
17. polymer flooding 聚合物驱油
18. viscous *adj.* 黏性的
19. surfactant *n.* 表面活性剂
20. immiscible *adj.* 不相混的
21. emulsion *n.* 悬浊液
22. surface tension 表面张力
23. caustic soda 苛性钠
24. API gravity API 重度(美国石油学会制定的一种衡量石油及石油产品相对于水的轻重的指标)

25. thermal recovery processes　热力采油
26. steam flooding　蒸汽驱
27. lateral permeability　横向渗透率
28. huff and puff　蒸汽吞吐
29. in-situ combustion　火烧油层
30. miscible hydrocarbon displacement　烃类混相驱
31. microbial EOR　微生物强化采油

Lesson 39 ▸ Reservoir Heterogeneity

It has been proposed that most reservoirs are laid down in a body of water by a long-term process, spanning a variety of depositional environments, in both time and space. As a result of subsequent physical and chemical reorganization, such as **compaction**, solution, dolomitization and cementation, the reservoir characteristics are further changed. Thus the heterogeneity of reservoirs is, for the most part, dependent upon the **depositional** environments and subsequent events.

The main **geologic characteristic** of all the physical rock properties that have a bearing on reservoir behavior when producing oil and gas is the extreme variability in such properties within the reservoir itself, both laterally and vertically, and within short distances. It is important to recognize that there are no **homogeneous** reservoirs, only varying degrees of heterogeneity.

The reservoir heterogeneity is then defined as a variation in **reservoir properties** as a function of space. Ideally, if the reservoir is homogeneous, measuring a reservoir property at any location will allow us to fully describe the reservoir. The task of **reservoir description** is very simple for homogeneous reservoirs. On the other hand, if the reservoir is heterogeneous, the reservoir properties vary as a function of a spatial location. These properties may include **permeability**, **porosity**, thickness, saturation, **faults** and **fractures**, rock facies and rock characteristics. For a proper reservoir description, we need to predict the variation in these reservoir properties as a function of spatial locations. There are essentially two types of heterogeneity:

(1) Vertical heterogeneity;

(2) Areal heterogeneity.

Geostatistical methods are used extensively in the **petroleum industry** to quantitatively describe the two types of the reservoir heterogeneity. It is obvious that the reservoir may be nonuniform in all intensive properties such as permeability, porosity, **wettability**, and connate water **saturation**. We will discuss heterogeneity of the reservoir in terms of permeability.

One of the first problems encountered by the production geology in predicting or interpreting fluid displacement behavior during **secondary recovery** and enhanced **oil recovery processes** is that of organizing and using the large amount of data available from **core analysis**. Permeabilities pose particular problems in organization because they usually vary by more than an order of magnitude between different **strata**. The engineer must be able to:

(1) Describe the degree of the vertical heterogeneity in mathematical terms;

(2) Describe and define the proper permeability **stratification** of the **pay zone**. This task is commonly called the **zoning or layering** problem.

It is appropriate to be able to describe the degree of heterogeneity within a particular system in quantitative terms. The degree of homogeneity of a reservoir property is a number that characterizes the departure from uniformity or constancy of that particular measured property through the thickness of reservoir. A formation is said to have a uniformity coefficient of zero in a specified property when that property is constant throughout the formation thickness. A completely heterogeneous formation has a **uniformity coefficient** of unity. Between the two extremes, formations have uniformity coefficients comprised between zero and one. The following are the two most widely used descriptors of the vertical heterogeneity of the formation:

(1) Dykstra-Parsons permeability variation V;

(2) **Lorenz coefficient** L.

Dykstra and Parsons introduced the concept of the permeability variation coefficient V which is a statistical measure of non-uniformity of a set of data. It is generally applied to the property of permeability but can be extended to treat other rock properties. It is generally recognized that the permeability data are log-normally distributed. That is, the geologic processes that create permeability in reservoir rocks appear to leave permeabilities distributed around the geometric mean. Dykstra and Parsons recognized this feature and introduced the **permeability variation** that characterizes a particular distribution. Dykstra-Parsons technique is an empirical way to assess reservoir heterogeneity.

Schmalz and Rahme introduced a single parameter that describes the degree of heterogeneity within a pay zone section. The term is called Lorenz coefficient and varies between zero, for a completely homogeneous system, to one for a completely heterogeneous system.

Since the early stages of oil production, engineers have recognized that most reservoirs vary in permeability and other rock properties in the **lateral direction**. To understand and predict the behavior of an underground reservoir, one must have as accurate and detailed knowledge as possible of the subsurface. Indeed, water and gas displacement is conditioned by the **storage geometry** (structural shape, thickness of strata) and the local values of the physical parameters (variable from one point to another) characteristic of the **porous rock**. Hence, prediction accuracy is closely related to the detail in which the reservoir is described.

Johnson and co-workers devised a well testing procedure, called pulse testing, to generate rock properties data between wells. In this procedure, a series of producing rate changes or pluses is made at one well with the response being measured at adjacent wells. The technique provides a measure of the formation flow capacity and storage capacity. The most difficult reservoir properties to define usually are the level and distribution of permeability. They are more variable than porosity and more difficult to measure. Yet an adequate knowledge of permeability distribution is critical to the prediction of reservoir depletion by any recovery process. A variety of geostatistical estimation techniques have been developed in an attempt to describe accurately the spatial distribution of rock properties. The concept of spatial continuity suggests that data points close to one another are more likely to be similar than are data points farther apart from one another. One of the best geostatistical tools to represent this continuity is a visual map showing a data set value with regard to its location. Automatic or computer contouring and griding is used to prepare these maps. These methods involve interpolating between known data points, such as elevation or permeability, and extrapolating beyond these known data values. These rock properties are commonly called regionalized variables. These variables usually have the following contradictory characteristics:

(1) A random characteristic showing erratic behavior from point to point;

(2) A structural characteristic reflecting the connections among data points.

New Words and Phrases

1. reservoir *n.* 储层
2. heterogeneity *n.* 非均质性
3. compaction *n.* 压实
4. depositional *adj.* 沉积作用的
5. cementation *n.* 胶结
6. homogeneous *adj.* 各向同性
7. porosity *n.* 孔隙性
8. permeability *n.* 渗透率
9. faults *n.* 断层
10. fracture *n.* 裂缝
11. wettability *adj.* 湿度
12. stratification *n.* 成层
13. strata *n.* 地层
14. saturation *n.* 饱和度
15. geologic characteristic 地质特征
16. reservoir properties 储层属性
17. reservoir description 油藏描述
18. geostatistical methods 地质统计学方法
19. petroleum industry 石油工业
20. secondary recovery 二次采油
21. core analysis 岩心分析
22. oil recovery processes 采油工艺
23. pay zone 产油层
24. zoning or layering 分区或分层
25. uniformity coefficient 均一性系数
26. Lorenz coefficient 洛伦茨系数
27. lateral direction 横向
28. storage geometry 储集空间
29. porous rock 多孔岩石
30. permeability variation 渗透率变异

Lesson 40 ▸ Reservoir Characterization

The volume of information that is being generated and made publicly available about oil and gas reservoirs is increasing at an exponential rate, as is most "knowledge". The "information age" applies equally to oil and gas **exploration** and development as to other global issues.

Partly because of the volume and the nature of available information, and the lessons learned and discussed from specific projects, the field of reservoir characterization is approaching a healthy level of maturity. Not many years ago, being assigned to evaluate a reservoir was considered a less desirable job for an exploration geologist. As exploration declined during the 1980s, **stratigraphers**, who had been accustomed to doing exploration evaluations, turned to describing reservoirs in order to enhance their employment capabilities in their chosen field. Exploration **geophysicists** also found a niche in reservoir development. Biostratigraphers and geochemists, among others, later found that their skills were applicable to reservoir characterization.

Even the **American Association of Petroleum Geologists** (AAPG) has now recognized the need for better balance between exploration and production. President P. J. F. Gratton stated: "… the growth of hydrocarbon recovery technology as a supplement and/or substitute for our traditional focus on discovery technology requires our attention and response."

Today, the field of reservoir characterization routinely involves disciplines of geology, geophysics, **petrophysics**, **petroleum engineering**, **geochemistry**, biostratigraphy, geostatistics, and computer science. Even behavioral science must be included in this list, because people in the different disciplines do not think or act similarly and sometimes must be encouraged to work together in a team setting. A popular quote (of unknown origin) is very appropriate to the oil and gas industry: Two stonecutters were asked what they were doing. The first said, "I'm cutting this stone into blocks." The second replied, "I'm on a team that's building a cathedral."

Different disciplines even use their own technical languages, so that communication is sometimes lacking and costly mistakes are made. An example is the term "**deep water**", which a geologist uses in the context of the **deposition** of **sediments** in water depths beneath storm **wave base** (**slope** and **basinal** depths). A drilling engineer refers to deep water in the context of drilling an **offshore** well in present-day water depths greater than 500m (1,500ft) above the mudline (seafloor).

Visualization technologies and equipment, introduced in earnest in the mid-1990s and now routinely used in all large and many midsize companies, have provided an effective means of breaking down the communication barriers among the disciplines. In part, this is due to the greater willingness of young entrants into the petroleum industry–geoscientists who were raised in an age of home and school computers and video arcades–to seek out, be comfortable with, and use computers for most tasks. Although we all acknowledge the advances that have been made in oil and gas exploration and development as a result of computers, there are a number of occasions of computer overuse (i. e., using the computer instead of knowledge to attempt to solve a problem). University professors are sometimes chided for producing "nintendo geologists" (a term first introduced to me by W. Camp, a longtime petroleum-industry geologist).

Thus, the field of reservoir characterization is quite comprehensive and challenging. In fact, definitions of reservoir characterization now vary according to the technologies available for characterization and the skills of the technologists. A rather vague definition that I like (perhaps because it is vague) has been provided by Halderson and Damsleth: "The principal goal of reservoir characterization is to outsmart nature to obtain **higher recoveries** with fewer wells in better positions at minimum cost through optimization."

The term reservoir characterization is frequently applied by geologists and engineers to the activities or results of qualifying and locating heterogeneities in reservoirs. By characterization is meant the description of the internal fabric of the rocks to produce an overall picture, or model, that details the nature of the reservoir. This model is then used for planning the extraction of the hydrocarbons. The reservoir description is compiled from detailed analysis of the **subsurface data** collected by means of core, **wireline logs**, **seismic survey**, and possibly includes pressure and production data.

However, **the term reservoir characterization** embraces more than just description, since the geological model must of necessity cover the entire volume of the reservoir, not just that penetrated by wells. Many workers fail to consider how much of their geological model is actually description from observations, and how much is prediction, in other words extension of the description into **unsampled regions**. The effective **spatial resolution** of seismic data, even 3D seismic, is rarely better than 10m, and usually is worse. This resolution is totally unsatisfactory for the kind of reservoir description now needed, though it may help with gross architecture. So the description has to rely on **borehole data**. However, even for highly developed fields, boreholes actually sample a minuscule proportion of the rock.

We must take some techniques in order to depict a reservoir appropriately. This text provides an overview of the common techniques for characterizing oil and gas reservoirs. The techniques can be subdivided into those that measure static reservoir properties and those that measure dynamic reservoir properties.

Static reservoir properties are those rock and fluid properties that normally do not change during the life of a field. They are the result of primary depositional processes coupled with postdepositional burial, **diagenesis**, and tectonics. Static properties include: stratigraphy, geometry, size, lithologies, structure, initial porosity and permeability, temperature.

Dynamic properties are those that do change significantly during the life of a field. For example, fluid saturations, compositions, and contacts, as well as reservoir pressure, change as the field is produced. Porosity and permeability can change as the reservoir pressure changes over time or as injected fluids react with formation minerals (either to precipitate new **minerals** that fill pore spaces or to dissolve minerals and thereby provide new pore spaces). Dynamic properties include: fluid saturations; fluid contacts; production and fluid-flow rates; pressure; fluid compositions, including gas-to-oil ratio (GOR) and water-to-oil ratio (WOR); **acoustic** (seismic) **properties**.

Acoustic properties, which are measured and documented as seismic attributes, are dependent upon porosity, fluid type and content, and the nature of the reservoir rock. Seismic attributes are dynamic, because fluid type and content change during oil and gas production. By comparing seismic attributes at different times in the life of a field, it is possible to indirectly measure fluid movement in the reservoir.

New Words and Phrases

1. characterization *n.* 描述
2. exploration *n.* 勘探
3. stratigrapher *n.* 地层学家
4. geophysics *n.* 地球物理学
5. geochemistry *n.* 地球化学
6. petrophysics *n.* 岩石物理学
7. deposition *n.* 沉积作用
8. sediment *n.* 沉积物
9. basinal *n.* 盆地
10. slope *n.* 斜坡，倾角
11. seismic *adj.* 地震的
12. diagenesis *n.* 岩化作用
13. minerals *n.* 矿物，矿石
14. offshore *adj.* 近海的
15. petroleum engineering 石油工程
16. deep water 深水
17. wave base 波基面，浪基面
18. high recovery 高采收率
19. subsurface data 地下数据
20. seismic survey 地震勘探
21. wireline logs 电缆测井
22. unsampled regions 未采样地区
23. spatial resolution 空间分辨率
24. borehole data 钻井数据
25. pore spaces 孔隙空间
26. acoustic property 声学特征
27. the term reservoir characterization 油藏描述术语
28. American Association of Petroleum Geologists (AAPG) 美国石油地质学家协会

Lesson 41 ▸ Modelling with Deep Learning

Building **geologically** realistic **facies models** based on **sparse measurements and interpretations** at wells is essential for field development and reservoir management. The modeling process involves predicting the spatial distribution of **sedimentary facies** over a wide **geographical area** given measurements at a few locations. Several tools exist for creating geological and petrophysical property models, one of the most important being **geostatistics**. Early **geostatistical** algorithms mainly used spatial linear interpolation or performed simulation of geological attributes by assuming that these attributes follow **Gaussian distributions**. This linear interpolation is based on a concept called a "**variogram**" that measures spatial continuity of the variable, such as porosity and geological facies. Most **geological patterns**, however, are non-Gaussian and highly nonlinear. To overcome these limitations, a new geostatistical approach called **multipoint statistics** was developed to simulate complex geological patterns based on a training image.

Recent research on applying **deep machine learning** to reservoir modeling using methods called **generative adversarial networks** (GANs), becomes more active. GANs have many potential applications, one of the most recognized being for image generation, where they have been used to generate artificial photorealistic facial images that are truly **indistinguishable** from actual pictures of real people. GANs are generative models composed of a generator, G, and discriminator, D, each parametrized by a separate neural network. G is trained to map a **latent vector** Z into an image X, while D is trained to map an image X to the probability of it being real versus having been generated. The networks are trained adversarially by optimizing the loss function [Eq. (1)]:

$$ {}_{G}\min_{D}\max[E_{X\sim P_{\text{data}}(X)}\cdot\ln D(X)]+E_{Z\sim P(Z)}\cdot\ln[1-D(G(Z))] \quad (1) $$

After training on data, drawn from a distribution P_{data}, G will be able to generate samples like those from P_{data}, by sampling $Z\sim P(Z)$ and mapping $X=G(Z)$. Training GANs is equivalent to optimizing a two-player game using a minimax objective function in which the **discriminator** aims to maximize reward by increasing the likelihood of correctly distinguishing real images from fake ones. Meanwhile, the generator attempts to reduce the risk that the generated images are recognized by the discriminator as being fake. Both G and D are trained alternatively, and the training process continues until it reaches an **equilibrium**. In other words, each player cannot improve itself, leading to a situation where the discriminator encounters difficulty in telling the difference between a true image and a fake one from the generator.

The generator creates fake images by starting with a **noise vector**, Z, in one-dimensional latent space, whose distribution is normal. The dimension of the latent space is usually low, such as 100 for 2D images and 200 for 3D models in our studies. This latent space can be considered as an embedded representation of the complex features or patterns from the true images in a much lower dimension. Once we have a trained GAN, the generator then can perform the prediction by generating new samples that resemble the true images (**training images**) but does not replicate them. Any noise vector drawn from the latent space can be used by the generator to map to a realistic-looking image. The prediction process is fast because of the reuse of the network parameters from the trained GAN, without the need for retraining. Figure 41.1 shows a GAN and illustrates the major components and their relationship involving two **adversarial networks**: a discriminator network against a generator network.

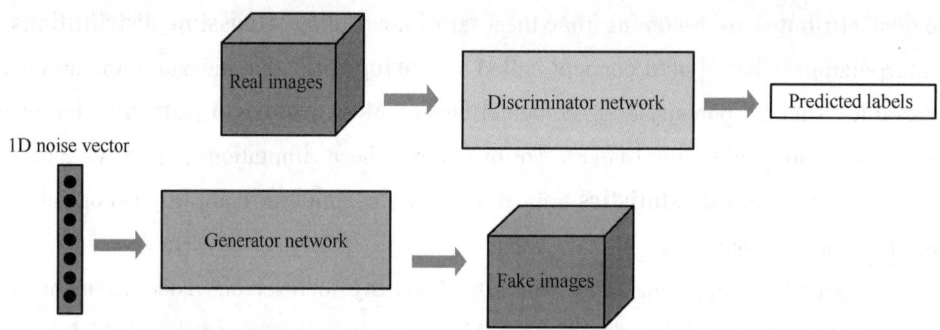

Figure 41.1 Schematic of a simple GAN

GANs is used to build geological facies models by feeding the networks with training images that are deemed to be suitable digital representations of conceptual geological models. The ability of 2D GANs is evaluated to generate unconditional fluvial samples or realizations. The fluvial training images are created using **object-based models** with varying channel widths and orientations. Figure 41.2 shows 16 of 15,000 fluvial samples generated (top of figure) with

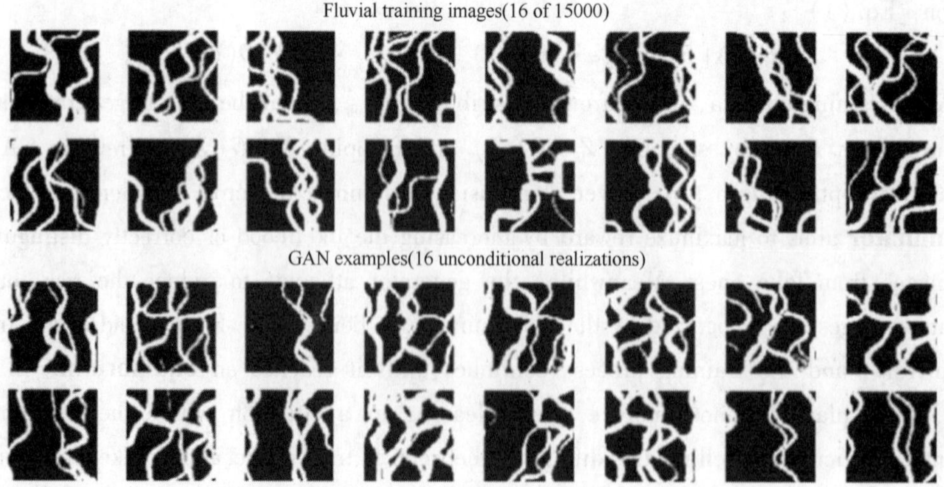

Figure 41.2 Fluvial training image examples (top) in 2D and unconditional samples or realizations generated by GANs (bottom)

simple binary facies (white: sand; black: background). The proportion of sand in all these training images is around 25%. The lower part of Figure 41.2 presents 16 generated unconditional fluvial models using GANs. It suggests that GANs can generate realistic fluvial images that are indistinguishable from the training images, specifically in the reproduction of channel width, **connectivity, and orientation**.

This method appears to be superior to existing geological modeling tools in several aspects. Firstly, it can generate realistic geological realizations with a wide range of implicit uncertainty by capturing a distribution of architectures and patterns, as opposed to a single training image used in MPS. Furthermore, it can be conditioned on a much larger number of known well measurements and interpretations than existing geostatistical simulation algorithms, while still generating realistic samples. Most importantly, it can capture and generate nonstationary facies patterns directly from nonstationary training images.

New words and Phrases

1. geologically facies model 地质相模型
2. sparse measurements and interpretations 稀疏的测点与解释
3. sedimentary facies 沉积相
4. exploration geologist 勘探地质学家
5. geographical area 地理区域
6. geostatistics *n.* 地质统计学
7. algorithm *n.* 算法
8. Gaussian distribution 高斯分布
9. variogram *n.* 变差函数
10. geological pattern 地质模式
11. multipoint statistics 多点统计学
12. deep machine learning 深度机器学习
13. generative adversarial network 生成性对抗网络方法
14. indistinguishable *adj.* 难以区分的
15. latent vector 潜在向量
16. discriminator *n.* 鉴别器
17. equilibrium *n.* 平衡
18. noise vector 噪声矢量
19. the training image 训练图像
20. adversarial network 对抗网络
21. object-based model 基于对象的模型
22. connectivity and orientation 连通性和方向性

Module 5

Progress of Energy Exploration

05

Lesson 42 ▸ Unconventional Petroleum Resources

Lesson 42

There is no formal definition of "unconventional resources" despite the fact that unconventional resources are the most active petroleum play in North America. Some workers defined unconventional resources using purely a permeability threshold (< 0. 1mD) . Yet, coal bed methane plays are considered unconventional and many have permeabilities exceeding 1mD over large portions of the fairway. Other workers have defined unconventional resources based on an interpretation of the petroleum system and have stated that unconventional resources are "continuous" or "basin centered" and lack traditional traps (Figure 42. 1) . While some have restricted the term to product type (i. e. unconventional gas), many shale and tight sand plays have gas, wet gas, and oil fairways and all can be considered unconventional. Heavy oil and oil sands are also unconventional resources and many of these deposits are in reservoirs with permeability exceeding 500 nD. Thus, unconventional resources include both low and high permeability reservoirs with both low and high viscosity fluids.

Unconventional Oil

A large share of the world's remaining oil resources is classified as unconventional. As there is no universally agreed definition of conventional or unconventional oil, several definitions are in current use to differentiate between them. Roughly speaking, any source of hydrocarbons that requires production technologies significantly different from the mainstream in currently exploited reservoirs is described as unconventional. However, this is obviously an imprecise and time - dependent definition; advances in technology, changes in the economics, new environmental requirements or energy policy incentives, can shift the demarcation line between

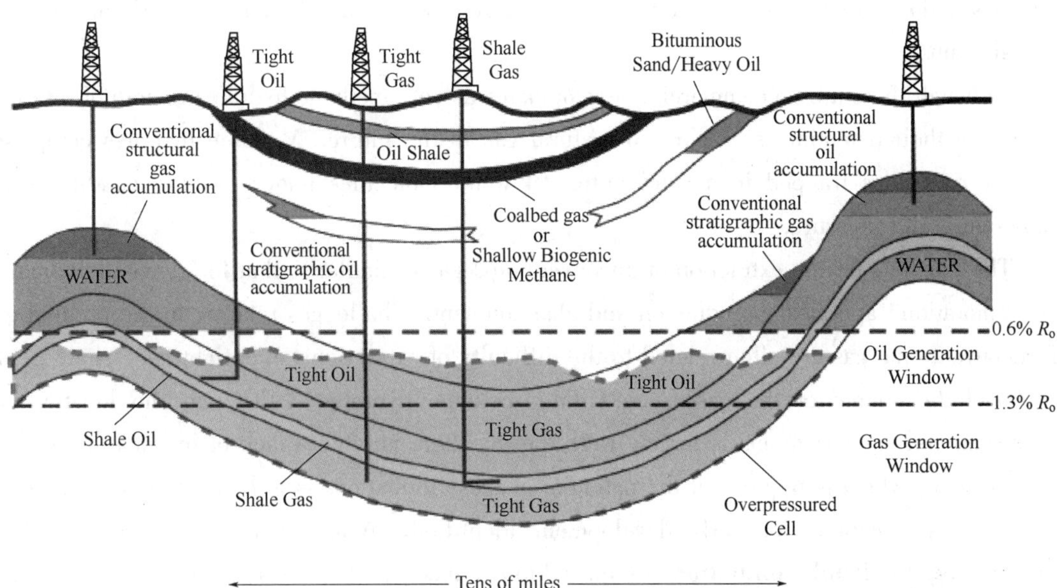

Figure 42. 1 A geological cross section showing different types of hydrocarbon accumulations (Sonnenberg and Meckel, 2016)

conventional and unconventional resources. In the long term, unconventional heavy oils may well become the norm rather than the exception.

Unconventional oil is generally accepted as having a high viscosity or is complex to extract; it usually includes **kerogen shale**, oil-sands (and natural bitumen), **light tight oil** (LTO), and oil derived from coal, gas and biomass. Natural gas liquids (NGLs) are occasionally included within this category, though more often as conventional. Although it is often more costly to produce, unconventional oil will almost certainly make an increasing contribution to future oil production.

Kerogen shale, also referred to as oil shale, generally refers to any sedimentary rock that contains kerogen, from which oil (or kerogen oil) may be produced by heating the kerogen. Oil sands contain a dense and extremely viscous form of petroleum, technically referred to as bitumen. Light tight oil (LTO) refers to light crude oil trapped in low permeability, low porosity shale, mudstone, siltstone, sandstone and carbonates formations.

Unconventional Gas

Unconventional gas is generally defined as natural gas that cannot be produced economically by using conventional technology.

The three common types of unconventional gas are tight gas, shale gas and coal-bed methane (CBM), though methane hydrates are often included. Tight gas refers to natural gas trapped in sandstone or limestone formations that exhibit very low permeability and low porosity; such formations may also contain condensate. Shale gas refers to natural gas trapped in organic-rich rocks, dominated by shale. CBM refers to natural gas adsorbed onto the matrix of the coal in coal seams.

Although there are no commercial developments as yet, methane hydrates remain of interest because of their potential as a source of natural gas in the future. Methane hydrates comprise methane molecules trapped in a solid lattice of water molecules under specific conditions of temperature and pressure.

The lifecycle for the extraction of unconventional gas is similar to that for conventional gas, i. e. exploration, appraisal, production and abandonment. Challenges relating to the production of unconventional gas are often related to the difficulty of extraction (e. g. tight gas, shale gas) and/or of the source of the gas (e. g. CBM, methane hydrates). Though the location of unconventional gas formations is largely well known, there is often a lack of information on the local geology, which is required if the potential of these formations is to be accurately evaluated and the best locations for early development identified. Where source rocks exhibit low permeability, **artificial stimulation** is required to enhance gas flow rates, which raises the costs of production. Production of unconventional gas is thus quite dependent on the extant (or future) price of natural gas.

Unconventional Tight Petroleum Systems

The driving forces for conventional accumulations (structural or stratigraphic traps) are forces of buoyancy which are due to differences in densities of hydrocarbons and water. In contrast, the **driving forces** for unconventional tight accumulations are forces of expulsion which are produced by high pressures. That is an enormous difference and creates unconventional petroleum systems that are characterized by very different and distinctive characteristics.

The force of **expulsion pressures** created by the significant increase in volume when any of the three main kerogen types is converted to hydrocarbons. At those conversion times in the burial history, the rocks are already sufficiently tight so the large volumes of generated hydrocarbons cannot efficiently escape through the existing tight pore system thus creating a permeability bottleneck that produces an **overpressured compartment** over a large area corresponding to the proper thermal oil and gas maturities for that basin. The forces initially created in these source rocks can only go limited distances into adjacent tight reservoirs (clastics or carbonates) above or below the source. The exact distance will vary depending on the pressure increase, matrix permeability, and fractures of that specific tight reservoir system. In general, the distances are small, in the orders of 10s to 100s of feet for oil and larger for more mobile gas systems. Those exact distance numbers are subject to ongoing investigations. Because the system is a pore throat bottleneck with very little or minimum lateral migration, the type of hydrocarbons are closely tied to the thermal maturity required to generate those hydrocarbons.

The play concept begins with two important geochemical considerations: (1) where are the source rocks and what are the kerogen types and organic richness (TOC)? (2) where are they mature for oil, condensate, and gas in the basin? These parameters will very quickly define the **fairway for the play**. Then one has to add the critical information on the reservoirs themselves: composition (brittleness), thickness, and reservoir quality (matrix porosity and permeability). In summary, these tight unconventional petroleum systems are dynamic and create a regionally **inverted petroleum system** with water over oil over condensate over gas for source rocks with Type Ⅰ or Ⅱ kerogen types.

New Words and Phrases

1. kerogen shale　干酪根页岩,油页岩
2. light tight oil　轻质致密油
3. artificial stimulation　人工改造
4. driving force　驱动力
5. expulsion force　排驱力
6. overpressured compartment　超压箱
7. fairway for the play　成藏组合区带
8. inverted petroleum system　反转含油气系统

Lesson 43 ▸ Oil Sands and Heavy Oil

Oil sands (natural bitumen) and **heavy oils** are conventional oils that have been subsequently degraded. They differ from light oil on the basis of their high viscosity, high density, and significant content of asphaltenes (rich in nitrogen, sulphur, oxygen and heavy metal contaminants). Heavy oils have API gravities of less than 22°, and viscosities of greater than 100 cP, while extra–heavy oil is that portion with API gravities of less than 10°. Natural bitumen is chemically similar but is even more viscous (>10,000cP).

These oils result from the same processes of petroleum generation, expulsion, migration, entrapment and alteration described elsewhere in this chapter. In this sense, they represent conventional petroleum systems. However, heavy oil and oil sands have experienced such an advanced state of alteration that they require unconventional methods of extraction. Volumes are extremely large, in numerous locations worldwide. Estimated original reserves of heavy oil and bitumen in known accumulations total 76 billion barrels and 250 billion barrels respectively at end 2005, as estimated by **the U. S. Geological Survey** . The statistics are dominated by two regions in particular. The largest natural bitumen deposit is the Canadian oil sands accumulation of Alberta, making up over 70% of the world's initial **recoverable bitumen reserves**. Over 90% of the world's **extra–heavy oil reserves** reside in the Orinoco Basin of eastern Venezuela.

Alberta Oil Sands

The Alberta oil sands represent a supercharged, laterally drained, low – impedance petroleum system. They have been a source of energy in Canada since the 1990s, and crude bitumen production now exceeds conventional crude production in Alberta. By 2015, the government of Alberta expects oil sands production to account for 65% of total Canadian crude oil output, and 17% of total North American supply. Volumetric estimates of the oil in place in the Alberta oil sands are vast, almost 1. 7 trillion barrels. 170 billion barrels are considered to be recoverable with existing technology, placing Canada second only to Saudi Arabia in terms of **proven reserves**.

The oil sands deposits of Alberta occur in the Peace River, Athabasca, and Cold Lake regions, mainly in the fluvio estuarine deposits of the Lower Cretaceous Mannville Group. These sediments lie on a major sub–Cretaceous unconformity that erodes down into strata ranging from the Triassic to the Devonian. Oil sands also occur to a lesser extent in subcropping Devonian carbonates. In addition, there is heavy oil production from the Mannville in the Loydminster area.

The Mannville Formation sands form part of a thick sequence of deltaic, fluvial and swampland sediments that was deposited in a flexural foredeep basin produced by the loading of thrust sheets in Late Jurassic–Early Cretaceous times during the Columbian orogeny.

Sediment supply was from the eroding thrust sheets in the west, and from the Precambrian shield to the east. Further Late Cretaceous–Neogene shortening in the Cordilleran region during the Laramide orogeny supplied large quantities of clastics to the marginal foredeep, driving further flexural subsidence and sediment loading. This phase of basin evolution was the critical moment for the petroleum system, being responsible for the maturation of pre–Cretaceous and Lower Cretaceous source rocks and the long–distance eastward lateral migration of very large quantities of petroleum.

The precise origin of the oils has been the subject of considerable debate over many years, but **biomarker analysis** of oils and source rocks has established the source and migration history of the Western Canada sedimentary basin. It is clear that the Mannville oil sands are the result of alteration of conventional crudes that migrated large distances from source kitchens to the west. According to Creany et al., several pre–Cretaceous source rocks have contributed to the Mannville Group oil sands, but the most important is the Lower Jurassic Nordegg Formation.

The Nordegg Formation, although often less than 20m thick, comprises an extremely organic–rich source rock containing highly oil–prone Type Ⅱ kerogen, with TOCs of up to 27% and hydrogen indices of up to 600. The Nordegg is mature for oil generation over most of its area of occurrence, and overmature in the far west. A liquid hydrocarbon mass balance suggests that the huge oil sands in–place volumes can be realistically explained by the source rock quality, history and geological setting of the basin. The highly porous and permeable Mannville sands acted as a migration conduit for oil moving through the subcropping pre–Cretaceous rocks, while the Joli Fou shale provides a regional topseal. 2–D modelling studies indicate northeastward migration of both oil and gas towards the present–day location of the oil sands deposits.

Meteoric water influx into the sandy outcropping Mannville Formation is responsible for biodegradation of the conventional crudes into bitumen and heavy oil.

Orinoco Basin, Eastern Venezuela, Heavy Oils

The Orinoco Basin petroleum system is supercharged, laterally drained, and of low impedance. The U. S. Geological Survey estimates in their 2000 World Petroleum Assessment that the Orinoco oil belt assessment unit of the East Venezuela province contains undiscovered resources of over 500 billion barrels. It represents the largest single oil accumulation in the world. The heavy oils have a range of API gravities between 4°~16°, and viscosities of 2000~8000 cP.

The source rock for the heavy oils is the Upper Cretaceous Querecual Formation, the stratigraphic equivalent of the La Luna Formation. The reservoirs are the fluvio–deltaic sandstones of the lower Miocene Oficina Group.

As for the Alberta oil sands, the critical moment for the petroleum system was a **regional plate tectonic event**, which created a thrust belt and adjacent foreland basin and flexural forebulge. Initially, the Albian - Turonian Querecual Formation marine pelagic algal - rich **calcareous mudstone** source rocks were developed on a north-facing passive margin prior to Eocene collision. The progressive eastward oblique collision of the Caribbean plate with the passive margin of northern South America, commencing in the Eocene in the west and continuing to the present day, formed a **transpressional thrust belt**. Lithospheric flexure resulting from the tectonic loading formed an adjoining foredeep together with a flexural forebulge on the Guyana shield to the south. This area of uplifted shield became a source of sediment for rivers draining into the foredeep from the south, and was responsible for the deposition of the chief reservoirs for the heavy oil play. From Oligocene times on, Cretaceous (and possibly older) source rocks buried under the thrust sheets and foreland basin generated large quantities of crude oil that migrated several hundred kilometres to the south through Upper Cretaceous and Miocene carrier beds, until trapped in Miocene clastic rocks in fault traps and stratigraphic pinch-outs against the Guyana shield in the region of the forebulge.

Shallow depths of burial and meteoric water influx have contributed to biodegradation of the reservoired oils.

New Words and Phrases

1. oil sands 油砂
2. heavy oils 重质油
3. the U. S. Geological Survey 美国地质调查局
4. recoverable bitumen reserves 可采沥青储量
5. extra-heavy oil reserves 超重质油储量
6. proven reserves 探明储量
7. biomarker analysis 生物标志物分析
8. regional plate tectonic event 局部板块构造活动
9. calcareous mudstone 钙质泥岩，石灰质泥岩
10. transpressional thrust belt 走滑挤压逆冲带

Lesson 44 ▸ Natural Gas-hydrates

Gas-hydrates were first obtained by Priestley (1778) under laboratory conditions by bubbling SO_2 through 0℃ water at atmospheric pressure. Priestley was a gifted researcher of his time who discovered a number of gases, in particular oxygen, hydrogen, SO_2 and others; however, when describing the crystals he obtained, he did not name them hydrates.

About 33 years later, similar crystals of aqueous **chlorine clathrate** (Davy, 1811) were named hydrates of gas. Some scientists consider Davy to be the discoverer of gas-hydrates; however, Priestley was the first scientist to create gas-hydrates in a laboratory. The results by Davy did not draw the attention of contemporaries and the studies of hydrates did not gain serious development for almost a century.

The Markhinskaya well drilled in 1963 in Yakutiya to a depth of 1800m revealed a section of rock at 0℃ temperature at the 1450m depth, with permafrost ending at approximately 1200m depth. Comparing the conditions of that section of rock with hydrate formation conditions allowed scientists to formulate an idea of the possibility of the existence of gas-hydrate accumulations in the cooled layers. After a comprehensive international examination, the discovery of natural hydrates was recorded in the USSR State Register of scientific discoveries, as the following formulation: "Experimentally established was the previously unknown property of natural gases to form deposit in the solid gas-hydrate state in the Earth's crust at specific thermodynamic conditions."

Soon thereafter, a group of young geologists named Sapir, Ben'yaminovich and Beznosikov found the first gas-hydrate deposit in the Messoyakha field in the Transarctic, on the eastern border of West Siberia. Comprehensive geophysical and thermodynamic studies performed in the Messoyakha wells showed that gas in the hydrated state existed in the upper part of the deposit. The underlying part of the deposit contained gas in a free state. The Messoyakha field, with original reserves of about $30 \times 10^9 m^3$, was dwarfed by the giant Urengoy, Yamburg and Medvezhye fields in Siberia.

The discovery of natural gas-hydrates coincided with the peak of one of the energy crises in the world. The urgency of the study of gas-hydrates become more important as the energy prices increased in the 1970s.

Gas-hydrates are compounds in which the molecules of gas are trapped in crystalline cells consisting of water molecules retained by the energy of hydrogen bonds. Gas-hydrates can be stable over a wide range of pressures and temperatures; for example, **methane hydrates** are stable from 20nPa to 2GPa at temperatures from 70K to 350K. The morphology of hydrate crystals is very diverse and is determined by composition and conditions of crystal growth.

Some properties of hydrates are unique. For example, $1m^3$ of water may tie up $207m^3$ of methane to form $1.26m^3$ of solid hydrate, while without gas $1m^3$ of water freezes to form $1.09m^3$ of ice. The dissociation of hydrates by increasing temperature when the volume is held constant will be accompanied by a substantial increase in pressure. For methane hydrate formed at 26 bar and 0℃, it is possible to obtain up to 1600 bar pressure increase. Hydrate density depends on its composition, pressure and temperature. Depending on the composition of gas, the density of the hydrate will vary from $0.8g/cm^3$ to $1.2g/cm^3$.

Natural gas-hydrates are metastable minerals, where the formation and/or decomposition depends on pressure and temperature, composition of gas, salinity of water, and characteristics of the **porous medium** in which they are formed. Hydrate crystals in reservoir rock can be dispersed in the pore space without the destruction of pores, but in some cases, the rock will be affected.

Hydrates can be in the form of small nodules (5cm to 12cm) or in the form of the small lenses or even of pure layers that can be several meters thick.

Liberating gas from a GHD (gas-hydrate deposits) requires heating up the entire rock mass containing the gas-hydrate. The amount of energy needed will depend on the heat capacity of the hydrate, the heat capacity of **the hydrate saturated layers**, the specific amount of hydrate in the layers and the degree of supercooling that caused the formation of the deposit. Under certain conditions, the energy necessary for liberating the gas in the hydrate layer can exceed the value of the potential energy of gas that will be produced. Hydrates possess high acoustic conductivity and low electrical conductivity, which is used in effective methods of finding and evaluating a GHD. The decomposition of hydrate in the layer, especially under the conditions of water areas, can be accompanied by significant changes in the strength of the sediments containing the gas-hydrates. The experimentally specific values of the heat of the formation or decomposition of hydrates of various hydrocarbons in the temperature range of the melting of water are given below.

The mechanism of how gas-hydrate deposits are formed and where hydrates are located has been affected by numerous factors, such as: (1) thermodynamic regime in the region; (2) intensity of **generation and migration of hydrocarbons**; (3) composition of the gas; (4) degree of gas saturation and salinity of the reservoir water; (5) structure of the porous medium; (6) **lithologic characteristics** of the section; (7) **geothermal gradients** in the zone of hydrate formation and in the basement rocks; (8) phase state of hydrate formers.

The hydrate formation zone (HFZ) represents the thickness of sediments in which the pressure and the temperature correspond to the **thermodynamic conditions** of stable existence of gas-hydrates of a specified composition. These HFZs are found where the Earth is cool, such as the Arctic and deep water. With an increase in the salinity of water, the thickness of the HFZ decreases. The thickness and the temperature of the HFZ in the offshore strongly depend on the value of sea bottom temperatures and gradient in the sediments. With an increase in sea bottom temperatures, the size of the HFZ decreases. In the regions where **permafrost** exists, the thickness of sediment in which gas-hydrate deposits exist can reach 400m to 800m. The HFZ in the ocean is found in the deepwater shelf and the oceanic slope in depths of water 200m or deeper for the conditions of in polar oceans, and from 500m to 700m or deeper for the equatorial regions. The upper boundary of the HFZ offshore is located near the sea floor.

New Words and Phrases

1. natural gas-hydrates 天然气水合物
2. chlorine clathrate 氯包合物
3. methane hydrates 甲烷水合物
4. porous medium 多孔介质
5. the hydrate saturated layers 天然气水合物饱和层
6. generation and migration of hydrocarbons 烃类的生成与运移
7. lithologic characteristics 岩石学特性
8. geothermal gradients 地温梯度
9. thermodynamic conditions 热力学条件
10. permafrost *n.* 永久冻土

Lesson 45 ▸ Geothermal Energy

Geothermal energy is heat within the Earth. The word geothermal comes from the Greek words *geo* (earth) and *therme* (heat). Geothermal resources are the thermal energy that could reasonably be extracted at costs competitive with other forms of energy at some specified future time, this definition was given by Muffler and Cataldi in 1978.

Geothermal energy is a **renewable energy** source because heat is continuously produced inside the Earth. The Earth's crust has immense heat (thermal) energy stored over millions of years. There exists a huge temperature difference between the Earth's crust and the surface. The temperature difference is known as **geothermal gradient**. This energy is sufficient to melt rock. The molten rock, called magma, at times erupts through cracks on the Earth surface as volcanoes. Geothermal energy is converted to produce to electricity.

Geothermal Energy Comes from Deep inside the Earth

The **slow decay of radioactive particles** in the Earth's core, a process that happens in all rocks, produces geothermal energy.

The Earth has four major parts or layers (Figure 45.1):

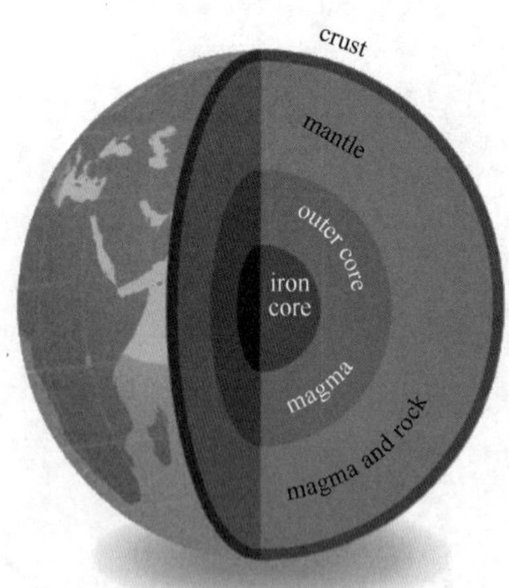

Figure 45.1 The Earth's interior

(1) An inner core of solid iron that is about 1,500 miles in diameter;

(2) An outer core of hot molten rock called magma that is about 1,500 miles thick;

(3) A mantle of magma and rock surrounding the outer core that is about 1,800 miles thick;

(4) A crust of solid rock that forms the continents and ocean floors that is 15 to 35 miles thick under the continents and 3 to 5 miles thick under the oceans.

Scientists have discovered that the temperature of the Earth's inner core is about 10,800 degrees Fahrenheit (°F), which is as hot as the surface of the Sun. Temperatures in the mantle range from about 392°F at the upper boundary with the Earth's crust to approximately 7,230°F at the mantle-core boundary.

The Earth's crust is broken into pieces called tectonic plates. Magma comes close to the Earth's surface near the edges of these plates, which is where many volcanoes occur. The **lava** that erupts from volcanoes is partly magma. Rocks and water absorb heat from **magma** deep underground. The rocks and water found deeper underground have the highest temperatures.

Geothermal Resources

Geothermal resources are generally confined to areas of the Earth's crust where heat flow higher than in surrounding areas heats the water contained in permeable rocks (reservoirs) at depth (Figure 45.2). The resources with the highest energy potential are mainly concentrated on the boundaries between plates, where visible geothermal activity frequently exists. By geothermal activity we mean hot springs, **fumaroles**, steam vents, and geysers. Active volcanoes are also a kind of geothermal activity, on a particularly and more spectacular large scale.

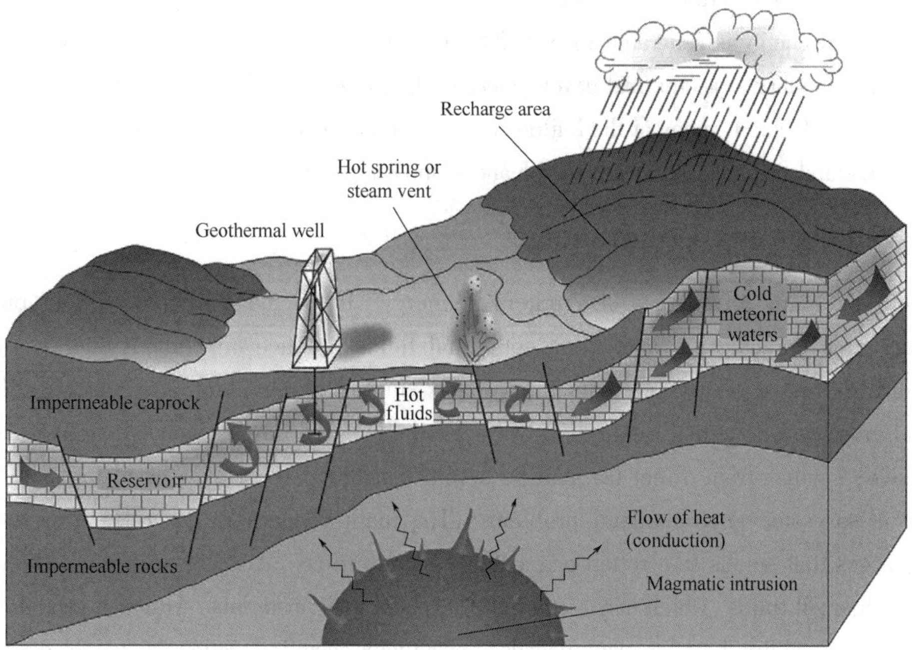

Figure 45.2 Geothermal resources

Geothermal activity in an area is certainly the first significant indication that subsurface rocks in the area are warmer than the norm. The local heat source could be a magma body at 600~1000℃, intruded within a few kilometres of the surface. However, geothermal fields can also form in regions unaffected by recent (Quaternary) shallow **magmatic intrusions**. The anomalous higher heat flow may be due to particular tectonic situations, for example to thinning of the continental crust, which implies the upwelling of the crust-mantle boundary and consequently higher temperatures at shallower depths.

However, we need more than a thermal anomaly to have a productive geothermal resource. We also need a reservoir, which is a sufficiently large body of permeable rocks at a depth accessible by drilling. This body of rock must contain large amounts of fluids, water or steam, which carry the heat to the surface. The reservoir is bounded by cooler rocks hydraulically connected to the hot reservoir by fractures and fissures, which provide channels for rainwater to penetrate underground. These cooler rocks crop out at the surface where they represent the so-called recharge areas of the **geothermal reservoir**. Thermal waters or steam are, in fact, mainly rainwater that infiltrates into the recharge areas at the surface and proceeds to depth, increasing in temperature while penetrating the hot rocks of the reservoir.

Production of Geothermal Energy

To produce geothermal-generated electricity, wells, sometimes a mile (1.6 kilometers) deep or more, are drilled into underground reservoirs to tap steam and very hot water that drive turbines linked to electricity generators. The first geothermally generated electricity was produced in Larderello, Italy, in 1904.

Geothermal energy is generated in over 20 countries. The United States is the world's largest producer, and the largest geothermal development in the world is the Geysers, north of San Francisco in California. In Iceland, many of the buildings and even swimming pools are heated with geothermal hot water. Iceland has at least 25 active volcanoes and many hot springs and geysers.

Advantages and Disadvantages

There are many advantages of geothermal energy. It can be extracted without burning a **fossil fuel** such as coal, gas, or oil. Geothermal fields produce only about one-sixth of the carbon dioxide that a relatively clean natural-gas-fueled power plant produces. Unlike solar and wind energy, geothermal energy is always available, 365 days a year. It is also relatively inexpensive; savings from direct use can be as much as 80 percent over fossil fuels.

But it has some environmental problems. The main concern is the release of **hydrogen sulfide**, a gas that smells like rotten egg at low concentrations. Another concern is the disposal of some geothermal fluids, which may contain low levels of toxic materials. Although geothermal sites are capable of providing heat for many decades, eventually specific locations may cool down.

Geothermal resources are reservoirs of hot water that exist at varying temperatures and depths below the Earth's surface. Mile-or-more-deep wells can be drilled into underground reservoirs to tap steam and very hot water that can be brought to the surface for use in a variety of applications, including electricity generation, direct use, and heating and cooling. In the United States, most geothermal reservoirs are located in the western states.

New Words and Phrases

1. geothermal energy 地热能
2. renewable energy 可再生能源
3. geothermal gradient 地温梯度
4. slow decay of radioactive particles 放射性粒子的缓慢衰变
5. lava *n.* 熔岩流

6. magma *n.* 岩浆
7. geothermal resources 地热资源
8. fumarole *n.* (火山地带的)喷气孔
9. magmatic intrusions 岩浆侵入体
10. geothermal reservoir 地热储层
11. fossil fuel 化石燃料
12. hydrogen sulfide 硫化氢

Lesson 46 ▸ Coal-seam Gas

Coal-seam gas or **coalbed methane** is a form of natural gas extracted from coal beds of both low rank and high rank coals. Coal-seam gas is an important energy resource because of the increasing demand for gas as a substitute for coal and oil in **electricity generation**. The coal-seam gas industry is relatively new in the history of the energy sector, with only around 30 years of commercial production. The United States is the top producer with 70%~80% of world production. Australia is the next highest producer with a 14-year production history followed by Canada, China and India. These countries extract natural gas from shallow to medium depth coal seams and also from coal seams in advance of or during coal mining.

The coal-seam gas refers to **adsorbed gas**, in this instance methane, in a near-liquid state inside the pores within the coal. Coal-seam gas is also called sweet gas because it is free of **hydrogen sulphide**. The **open fractures** in the coal can also contain free gas or can be saturated with water. Production of coal-seam gas is achieved by reducing the pressure within the coal seam that releases the **methane** from the coal in the form of gas. The methane is brought to the surface accompanied by the coal-seam gas water as the pressure is released.

Coal-seam gas is mainly a combination of methane (CH_4) and carbon dioxide and has very close methane percentage (95%) to that of natural gas (97%). In contrast, biogas produced from the digestion process has methane content around 68%, which makes it low energy content. Due to the decline in the natural gas resources and the low quality of **biogas**, coal-seam gas can be the energy resource solution for the next few decades. This can be true until the technology is able to enhance the production of biogas from **anaerobic digestion process** and reduce the cost of **purifying the gas**.

Coal-seam gas, as it is known in China, is becoming an increasingly important source of energy around the world. Many countries such as United States, Canada, and Australia are investing in the coal-seam gas industry. A rise in the cost of **conventional natural gas** and many other energy resources, along with a decline in these conventional resources and issues such as climate change have encouraged a global interest in **alternative sources** of energy like coal-seam gas. The estimated quantity of coal-seam gas worldwide is around 1.4×10^{14} m^3, it is clear that coal-seam gas is a significant source of energy.

The importance of coal-seam gas as an energy source, as an example in Australia, can be shown in Table 46.1. The table shows that coal-seam gas contribution in the energy production is higher than petroleum products (oil, condensate, and LPG) individually. It also represents **approximately** 11.5% of the total need for gas energy in Australia. Under current production rate coal-seam gas is expected to last for 100 years which is higher than petroleum products, black coal and conventional gases.

Table 46.1 Coal-seam gas as a source of energy in Australia compared to other sources

Energy source	Amount(10^{15}J)	Ratio of economic demonstrated reserves
Black coal	987,064	111
Brown coal	359,870	539
Oil	6,290	10
Condensate	12,691	45
Liquefied petroleum gas	4,399	42
Conventional gas	123,200	68
Coal seam methane	16,180	100
Uranium	685,440	141

The coal-seam gas development in China has increased significantly through the past decade. Until 2003 there were only 250 documented coal-seam gas wells, however, this number increased to 2,500 wells in 2008. The total coalbed methane resources present in China in Huaibei and Huainan coalfields exceed $1.4\times10^9 m^3$. Coalbed methane in China represents a very promising source of energy and some studies recognize the geological coalbed methane resource volume as third in the world behind the United States and Canada. There are nine major coalbed methane basins in China, their total reserve is $30.9\times10^{12} m^3$, which is 84% of the total resources in China. In recent years, natural gas demand in China has grown significantly due to the rapid and continuous growth of its economy.

As coal seams (and the gases within them) are formed over long times through coalification, there can be many **geological variations** that occur throughout a region. This can have significant impacts on the structure of coal seams, and therefore the production of gas and water, as there is a high degree of **spatial heterogeneity** (variations in coal seam thickness, character and **lateral continuity**) across the field. The **keen demand** and rapid mining are draining gas resource from the easy-recovery coalbeds, leaving a great amount of reserves in the deep and **low permeability** coal seams.

Heat-injection to stimulate coal seam has gained great attention as a **supplementary enhancement method** for gas production from deep coal in recent years. Heat-injection through the **hydraulic fractures** might increases the production efficiency from shale formation by improving the gas desorption rate, especially for the long-term gas production. **Microwave and radio frequency** could also heat gas reservoir, it has essential distinctions in the performances of chemical reaction. **Injecting geothermal resource** into coal seam is one kind of thermal stimulation patterns to enhance gas recovery. Lastly, the heated nitrogen and carbon dioxide are proposed as potential means to drive or displace the free and adsorbed gas from coal seams.

New words and Phrases

1. coal-seam gas 煤层气
2. coalbed methane 煤层气
3. electricity generation 发电量
4. adsorbed gas 吸附气
5. hydrogen sulphide 硫化氢
6. open fractures 开放性裂缝
7. methane *n*. 甲烷
8. biogas *n*. 沼气
9. anaerobic digestion process 厌氧消化工艺
10. purifying the gas 净化气体
11. conventional natural gas 常规天然气
12. alternative source 替代资源
13. approximately *adv*. 大约
14. geological variation 地质变化
15. spatial heterogeneity 空间非均质性
16. lateral continuity 侧向连续性
17. keen demand 强烈的需求
18. low permeability 低渗透率
19. heat-injection *n*. 热注入
20. supplementary enhancement method 补充增强方法
21. hydraulic fracture 水力裂缝
22. microwave and radio frequency 微波和射频
23. injecting geothermal resource 注入地热资源

Lesson 47 ▸ Tight-gas Sands

Tight-gas sands are natural extensions of conventional sandstone reservoirs with natural-gas accumulation but with permeability values less than 0.1mD, which may be as low as 0.000001mD. They also generally have lower effective porosity. It is estimated that globally 600 tcf of natural gas is recoverable. Because of the tight characteristics of this class of resources, large **hydraulic-fracture** stimulation treatments through **horizontal or multilateral wells** are necessary to be able to recover economic volumes of gas. Horizontal/multilateral wells combined with a large number of discrete hydraulic fractures increase the surface area in contact with the formation, resulting in reduced drawdown and higher-conductivity paths for production. These operational features, which are different from conventional reservoirs, require additional considerations to minimize risks and maximize economic efficiency.

Existence of low ranges of permeability brings on additional challenges to characterization and evaluation of both dynamic and static reservoir parameters from well logs, borehole/surface **microseismic surveys** and core samples. For example, pressure transient testing response is slower than conventional reservoirs, which reduces the practicality of these tests to obtain permeability estimation in a reasonable time. Thorough fracture treatment and extended flow/buildup periods are typically needed for realistic interpretations. Complex interplays between petrophysical properties create challenges for correlating them with core permeability. Multiple relationships between porosity and permeability may be needed for proper characterization. Porosity estimations from wireline logs may be **erroneous** due to the factors such as **shallow invasion**, presence of heavy minerals and **diagenetic factors**. Another example is the usage of **various correlations** that are employed to estimate permeability values from well log data. These **empirical equations** include a wide range of exponents and parameters for different types of reservoir rocks. The complexity and heterogeneity of tight-gas systems can lead to large errors in accurate predictions of permeability, thickness, porosity, and hydraulic fracture properties such as width, length and permeability, and at the end of the day, can result in **misleading interpretations**. Flow-model building process is **resource-demanding** due to such difficulties in **reservoir characterization**. Because of these reasons, the **history-matching process** for a tight-gas sand reservoir model also becomes lengthy and challenging.

An inverse reservoir-characterization model is developed that can be used to **address the aforementioned issues** by providing accurate estimations of some unknown reservoir and hydraulic-fracture characteristics. The presented study is a continuation and the inverse-looking component of the work presented by Kulga et al., in which horizontal well performance (future gas production defined with decline-curve coefficients) was estimated from reservoir and hydraulic-fracture characteristics. Similar to that study, the model is trained with a large dataset using an **artificial neural network** with an optimized design. The dataset is obtained by running a large number of scenarios using a representative numerical model for tight-gas sand reservoirs. **Performance characteristics** are represented with **decline-curve coefficients**, which allows us to generalize the production decline without depending on time. The structure of the presented model is desired to estimate reservoir and hydraulic-fracture characteristics, once controllable operational parameters, known initial conditions, and observed horizontal well performance, in the form of hyperbolic decline-curve coefficients, are input. This structure makes it an inverse model that can provide solutions to the inverse problem of reservoir characterization of tight-gas sand reservoirs. Therefore, the model serves as a **complementary tool** to the one presented by Kulga et al. by solving the much more challenging inverse problem. The efficiency in the computational time with regard to the **data-driven** nature of the model allows the practicing engineer to achieve the modeling objectives much quicker than a numerical model. In this way, **unavoidable uncertainty ranges** associated with unknown characteristics would also get reduced for further numerical modeling work.

Tight-gas systems are rather complex geological and petrophysical systems. Supplementary tools are generally needed to reduce the uncertainties related to unknown reservoir and hydraulic-fracture characteristics. Artificial-neural-network based models were successfully applied to various inverse problems in **petroleum engineering**, such as reservoir characterization and **operational optimization applications**, using both real-field data and synthetic or numerical-model based data. Examples for these successful applications include estimating fracture density by using production data, predicting well logs from seismic attributes, predicting well log facies, designing cyclic pressure pulsing in naturally fractured reservoirs, and designing lateral well configurations. A similar inverse model is used to solve complex problems for the characterization of tight-gas systems, especially for the following situations:

(1) To support a wide range of estimation problems encountered in well-logging, well testing, borehole or surface microseismic surveys and core sampling of complex tight-gas systems, when accurate petrophysical or fracture-design data do not exist.

(2) To obtain rapid solutions of inverse problems associated with computationally expensive simulation models of tight-gas systems (e. g. , during a history-matching study).

This methodology can be used to investigate many cases rapidly, and the methodology proposed is practical and computationally efficient. This is demonstrated with a case study by incorporating an observed range for production performance into **Monte Carlo simulation**. Probabilistic estimates of reservoir and hydraulic-fracture parameters are quantified and they are found to be in agreement with the pre-defined uncertainties in the numerical model. This characteristic of the tool can provide guidance to engineers before a more detailed history-matching study for a numerical model.

New words and Phrases

1. tight-gas *n.* 致密气
2. hydraulic-fracture *n.* 水力压裂
3. horizontal or multilateral wells 水平井或者丛式井
4. microseismic survey 微地震测量
5. erroneous *adj.* 错误的
6. shallow invasion 浅层侵入
7. diagenetic factor 成岩因素
8. various correlations 多种相关性
9. empirical equation 经验公式
10. misleading interpretation 误导性解释
11. resource-demanding *n.* 资源需求
12. reservoir characterization 储层表征
13. history-matching process 历史拟合过程
14. address aforementioned issues 解决上述问题
15. artificial neural network 人工神经网络
16. performance characteristic 性能特征
17. decline-curve coefficient 衰减曲线系数
18. complementary tool 补充工具
19. data-driven *n.* 数据驱动
20. unavoidable uncertainty range 不可避免的不确定度范围
21. petroleum engineering 石油工程
22. operational optimization application 操作优化应用
23. Monte Carlo simulation 蒙特卡罗模拟

Lesson 48 ▸ Shale Gas

What are Shale and Shale Gas?

Shale is a sedimentary rock, which was once deposited as mud (clay and silt) and is generally a combination of clay, silica (quartz), carbonate (calcite or dolomite), and organic material. Mainly shale is a composite of a large amount of kerogen, which is a mixture of organic compounds. As a primary composition, it has kerogen, quartz, clay, and carbonate. Uranium, iron, vanadium, nickel, and molybdenum are present as secondary components. From this rock, the shale hydrocarbons (liquid oil and gas) are extracted. Various shales exist; namely, black shale (dark) and light shale. Black shale has a rich content of **organic matters**, while light shale has much less content of organic matters, relatively. Black shale formations were buried under little or no presence of oxygen and this preserved the organic matters from decay. This organic matter may produce oil and gas through a heating process. Many shale formations in the USA are **black shale formations** which give natural gas via heating.

In the case of shale gas formations, natural gas is generated in the shale and remains within the shale. Consequently the shale is both source rock and reservoir rock. Grains and pores are smaller in shales as compared with tight and conventional gas formations. Shale gas is trapped within the pores of this sedimentary rock. Gas is normally stored through three ways in gas shales:

(1) **Free gas**: The gas is within the rock pores and natural fractures.

(2) **Adsorbed gas**: The gas is adsorbed on organic materials and clay.

(3) **Dissolved gas**: The gas is dissolved in the organic materials.

Over the past decade, the combination of **horizontal drilling** and **hydraulic fracturing** has allowed access to large volumes of shale gas that were previously uneconomical to produce. The production of natural gas from shale formations has rejuvenated the natural gas industry.

Important Parameters in the Shale Gas Context

1. Type of Shale

There are different types of shale in terms of material content and origin. The shales are marine or nonmarine. **Marine shales** have low clay content. They are high in **brittle minerals** such as quartz, feldspar, and carbonates. Hence, they respond better to hydraulic stimulation.

2. Depth

In general, the depth is directly related to the amount of natural and generated hydrocarbons deposited in the formation. For instance, gas as a biogenic gas is formed through

anaerobic microorganisms throughout the early stage of the burial process or/and thermogenic breakdown of kerogen at higher temperatures and depths. The typical depth for shale gases ranges from 1000m to 5000m. The shale formations shallower than 1000m normally experience lower gas concentrations and pressures, while regions with a depth greater than 5000m commonly have reduced permeability which is translated into higher costs for drilling activity and field development.

3. Adsorbed Gas

Adsorbed gas is the gas accumulated on the surface of a solid material, such as a grain of a reservoir rock, or more particularly the organic particles in a shale reservoir. Measurement of adsorbed and interstitial gas (existing in pore spaces) allows calculation of gas in place in a reservoir.

4. Organic Maturity

This is expressed in terms of **vitrinite reflectance** R_o. The range of 1.0% ~ 1.1% indicates that organic matter is adequately mature to generate gas. Generally, higher gas in-place resources should be produced by more mature organic matter.

5. Permeability

The permeability of any type of porous media is defined as the ability of fluid (i. e., gas, oil, water) flow due to a pressure difference through the porous system. Therefore, it implies the fluid transmissivity and storage features of a shale formation. Permeability of shales is generally very low ($< 10^{-3}$mD). Hence, artificial stimulations (particularly hydraulic fracturing) are needed to ease the hydrocarbon flow toward the well. If natural fractures exist in the shale formation, it is crucial to map the orientation and intensity of the open fractures. If the fractures are poorly cemented or open, stimulation will open these formerly created regions of weakness. In some circumstances, compactly cemented fractures can create barriers of fractures which are measured in millidarcies (mD).

6. Porosity

Porosity is the percentage of void space versus solid rock, which is the space where gas is potentially trapped. The porosity of shale reserves is normally lower than 10%.

7. Reservoir Thickness

This defines the vertical extent thickness of the productive portion of a reservoir. The formation thickness varies from one shale reserve to another shale reservoir. The typical range of the thickness is 2~5m.

8. Total Organic Content (TOC)

This is expressed as the total amount of organic material present in the rock (a percentage by weight). The higher the TOC, the higher is the potential for hydrocarbons (HCs) production. Typical values are equal to or greater than 1%. The TOC and thermal maturity of source rocks are assessed by means of lab analysis.

9. Thermal Maturity

This is the measure of the extent to which organic matters contained in the rock have been heated over time and potentially converted to liquid and/or gaseous HCs. The indicator for this measure is called vitrinite reflectance and has typical values ranging from 1% to 3%.

10. Viscosity

Viscosity is a measure of how easily oil will flow. Inside the reservoir, viscosity is measured in poises (P); it is normally measured in centistokes (cS) outside the reservoir.

11. Mineralogy

The mineral structure in shale formations is complicated. One ought to endeavor to obtain a few cores for an underlying assessment in a new region. Electron catch spectroscopy (ECS) logs give a good estimation of mineralogy, but not as the mineral (e. g., granular against cryptocrystalline), which plays a significant role in brittleness behavior. One effective strategy is to build a ternary outline of total aggregate carbonate, total aggregate mud, and quartz and guide it to elastic parameters (e. g., Young's modulus and Poisson's ratio), leading to a brittleness template. Such layouts can then be adjusted to production logs, microseismic event area, and production itself to evaluate the ductility or brittleness of the rock and how well the induced fractures have actuated it.

12. Fluid in Place

In general, fluid in place is determined using TOC, porosity, temperature, and pressure data for economical evaluation of the shale.

13. Free Gas Quantification

Adsorption phenomenon is considered as a more effective mechanism at gas storage under low pressures, whereas free gas signifies the main amount of gas at elevated pressures. The free gas percentage in shale gases varies in the range of 15%~80%, depending on gas saturation, porosity, and reservoir pressure. Hence, determination of free gas is essential to describe/characterize gas shales. Thus, the quantification of free gas is also necessary to characterize gas shale.

New words and Phrases

1. shale gas　页岩气
2. coalbed methane　煤层气
3. organic matters　有机质
4. black shale formations　黑色页岩层
5. free gas　游离气
6. adsorbed gas　吸附气
7. dissolved gas　溶解气
8. horizontal drilling　水平钻井
9. hydraulic fracturing　水力压裂
10. marine shales　海相页岩
11. brittle minerals　脆性矿物
12. anaerobic microorganisms　厌氧微生物
13. organic maturity　有机质成熟度
14. vitrinite reflectance　镜质组反射率

Lesson 49 ▸ Shale Gas Resources and Exploration

China: The Largest Potential World Shale Gas Source

China has seven **prospective shale gas basins**: the Sichuan, Tarim, Junggar, Songliao, Yangtze Platform, Jianghan, and Subei. In 2013, **the U.S. Energy Information Administration** (**EIA**) estimated China's **technically recoverable shale gas reserves** to be approximately 1,115 trillion cubic feet (tcf), the largest in the world. Estimates have varied, however, and in 2012, **China's Ministry of Land and Resources** (**MLR**) estimated technically recoverable reserves to be 883 tcf. In 2016, MLR announced that China's recoverable shale gas reserves—those that can be commercially produced—rose by 109 billion cubic meters (bcm) in 2015, bringing China's total shale gas reserves to 130bcm. In 2017, China's shale industry reportedly reached nearly 600 wells and 9 bcm of production, with output expected to nearly double to 17bcm by 2020.

Compared with the United States, the commercialization of shale gas in China has been challenging, due to the more complex geology, the presence of older strata, more complex tectonism, more complicated thermal histories, and higher levels of thermal maturity. Considerable difficulties in specific fields need to be overcome if future shale gas exploration is to be successful, including understanding accumulation mechanisms, laboratory testing, evaluation and forecasting of **sweet spots**, productivity analysis and prediction, hydraulic fracturing technology, geological steering, and reservoir stimulation.

Shale Gas Resources

Of the natural gas consumed in the United States in 2009, 87% was produced domestically; thus, the supply of natural gas is not as dependent on foreign producers as is the supply of crude oil, and the delivery system is less subject to interruption. The availability of large quantities of shale gas will further allow the United States to consume a predominantlydomestic supply of gas.

Shale Gas vs. Conventional Gas

Conventional gas reservoirs are created when natural gas migrates toward the Earth's surface from an organic-rich source formation into highly permeable reservoir rock, where it is trapped by an **overlying layer** of impermeable rock. In contrast, shale gas resources form within the organic-rich shale source rock. The low permeability of the shale greatly inhibits the gas from migrating to more permeable reservoir rocks. Without horizontal drilling and hydraulic fracturing, shale gas production would not be economically feasible because the natural gas would not flow from the formation at high enough rates to justify the cost of drilling.

Exploration and Drilling in Shale Gas and Oil Reserves

1. Introduction

In the past, hydrocarbons trapped in shale and tight reservoirs have been uneconomical to produce, as the return on investment using traditional drilling and production methods was too low. As a result of the advances in drilling technology, exploration and drilling in these reservoirs has increased, unlocking untapped resources. These technologies include **horizontal drilling**, **hydraulic fracturing**, and advanced **drilling fluids**, among others.

2. Exploration Techniques

There are several exploration techniques employed in the oil and gas industry for exploring for shale oil and gas. One of these techniques is **seismic surveying**. Seismic surveying uses large machinery at the surface that creates a vibration sending **seismic waves** down through the Earth. These seismic waves reverberate differently off the different ground layers and are recorded by **geophones** at the surface. Based on how the waves are reflected back and how long they take to return a 2D or 3D model of the Earth can be created.

A third seismic model can also be created and this is a 4D model. The 4D model incorporates seismic data or core samples taken over an extended period of time. This allows the changes in the rock formation over time to be examined. Using these seismic images potential traps and oil reservoirs can be determined based on the rock formations and rock types.

Another method of exploration, which is generally carried out after seismic surveying, is **geophysical well logs**. This is the method of actually drilling into the Earth's surface and removing a core sample which is then further analyzed in a lab. Analyzing core samples in a lab allows the prospector to identify any oil or gas trapped in the core sample as well as the key potential rock properties such as porosity, permeability, and wettability. This process could then further lead to the identification of a valuable reservoir.

3. Horizontal Drilling

Two major drilling techniques are used to produce shale gas. Horizontal drilling is used to provide greater access to the gas trapped deep in the **producing formation**. First, a vertical well is drilled to the targeted rock formation. At the desired depth, the **drill bit** is turned to bore a well that stretches through the reservoir horizontally, exposing the well to more of the producing shale.

4. Hydraulic Fracturing

Hydraulic fracturing (commonly called "fracking" or "hydrofracking") is a technique in which water, chemicals, and sand are pumped into the well to unlock the hydrocarbons trapped in shale formations by opening cracks (fractures) in the rock and allowing natural gas to flow from the shale into the well (Figure 49.1). When used in conjunction with horizontal drilling, hydraulic fracturing enables gas producers to extract shale gas at reasonable cost. Without these techniques, natural gas does not flow to the well rapidly, and commercial quantities cannot be produced from shale.

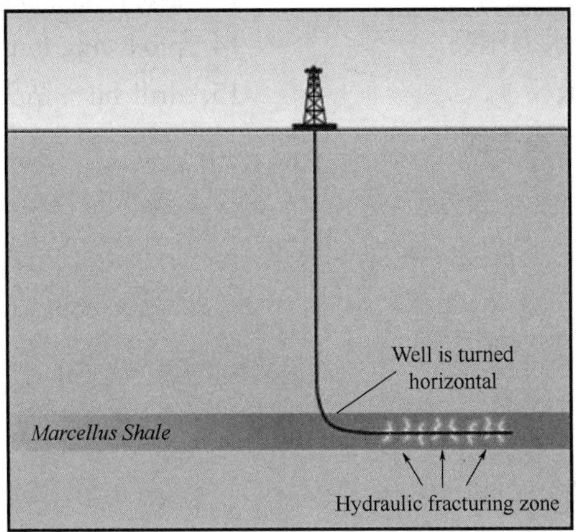

Figure 49.1 Diagram showing the concept of a horizontal well with hydraulic fracturing zone

Environmental Concerns

However, there are some potential environmental issues that are also associated with the production of shale gas. Shale gas drilling has significant water supply issues. The drilling and fracturing of wells require large amounts of water. In some areas of the country, significant use of water for shale gas production may affect the availability of water for other uses, and can affect aquatic habitats.

Drilling and fracturing also produce large amounts of wastewater, which may contain dissolved chemicals and other contaminants that require treatment before disposal or reuse. Because of the quantities of water used, and the complexities inherent in treating some of the chemicals used, wastewater treatment and disposal is an important and challenging issue.

If mismanaged, the hydraulic fracturing fluid can be released by spills, leaks, or various other exposure pathways. The use of potentially hazardous chemicals in the fracturing fluid means that any release of this fluid can result in the contamination of surrounding areas, including sources of drinking water, and can negatively impact natural habitats.

New words and Phrases

1. prospective shale gas basins 有远景的页岩气盆地
2. the U.S. Energy Information Administration (EIA) 美国能源信息管理局
3. technically recoverable shale gas reserves 技术可采页岩气储量
4. China's Ministry of Land and Resources (MLR) 中国国土资源部

5. sweet spots　甜点
6. overlying layer　上覆地层
7. horizontal drilling　水平钻井
8. hydraulic fracturing　水力压裂
9. drilling fluids　钻井液
10. seismic surveying　地震探测
11. seismic waves　地震波
12. geophones　*n.* 地震检波器
13. geophysical well logs　地球物理测井
14. producing formation　（生）产层
15. drill bit　钻头

References

[1] Ansley C F, et al. The Columbia encyclopedia. 6th edition. New York: Columbia University Press, 2008.

[2] Skinner B J, Park S J. The dynamic Earth: an introduction to physical geology. New Jersey: John Wiley&Sons Inc, 2003.

[3] Lu Hongbo. An outline of Earth sciences. Dongying: China University of Petroleum Press, 2006.

[4] Hancock P, Skinner B J. Sedimentology, sediments, and sedimentary rocks. Oxford: Oxford University Press, 2000.

[5] Selly R C. Applied sedimentology. London: Academic Press, 2000.

[6] Zimmerle W. Petroleum sedimentology. Springer Verlag: Kluwer Academic Publishers, 1995.

[7] Prothero D R, Schwab F. Sedimentary geology: an introduction to sedimentary rocks and stratigraphy. New York: W H Freeman, 2004.

[8] Prothero D R. Sedimentary geology. New York: W H Freeman, 2004.

[9] Richard A, Davis J R. Depositional systems: an introduction to sedimentology and stratigraphy. New Jersey: Prentice Hall, 1992.

[10] Reading H G. Sedimentary environments: processes, facies and stratigraphy. Blackwell science Ltd, 1996.

[11] Nichols G. sedimentology and stratigraphy. New Jersey: John Wiley&Sons, 2009.

[12] Mikhailov A Y. Structural geology and geological mapping. Moscow: Mir Publishers, 1987.

[13] Dercouet J H. Geological principles and methods. Vancouver: University of British Columbia, 1985.

[14] Pollard D D, Fletcher R C. Fundamentals of structural geology. New York: Cambridge University Press, 2005.

[15] Twiss R J, Moores E M. Structural geology. New York: W H Freeman, 2007.

[16] LeGrand H E. Drifting continents and shifting theories. New York: Cambridge University Press, 1988.

[17] Silverstein A, Silverstein V, Nunn L S, et al. Plate tectonics. Lerner Publishing Group Inc, 2009.

[18] Chilingar G V, Buryakovsky L A, et al. Geology and geochemistry of oil and gas. Amsterdam: Elsevier Inc, 2005.

[19] Hunt J M. Petroleum geochemistry and geology. New York: W H Freeman, 1995.

[20] Tissot B P, Welte D H. Petroleum formation and occurrence. Berlin: Spring-Verlag, 1984.

[21] Selley R C. Elements of petroleum geology. London: Academic Press, 1985.

[22] Magoon L B, Dow W G. The petroleum system from source to trap. AAPG, 1994.

[23] Otis R M, Schneiderman N. A process for evaluating exploration prospects. AAPG Bulletin, 1997, 81(7): 1087-1109.

[24] Beaumont E A, Foster N H. Exploring for oil and gas traps. American Association of Petroleum Geologists, 1999.

[25] Laudon R C. Principles of petroleum development geology. New Jersey: Prentice Hall, 1996.

[26] Ahmed T. Reservoir engineering handbook. 4th edition. Amsterdam: Elsevier Science & Technology, 2010.

[27] Slatt R M. Stratigraphic reservoir characterization for petroleum geologists geophysicist engineers. Amsterdam: Elsevier Inc, 2006.

[28] Lyons W C, Guo B, Graham R L, et al. Air and gas drilling field guide. 3rd edition. Amsterdam: Elsevier Inc, 2009.

[29] Ward C R. Coal geology and coal technology. Oxford: Blackwell Scientific Publications, Australia, 1984.

[30] Selley R C. An introductory to sedimentology. London: Academic Press, 1982.

[31] Galloway W E, Hobday D K. Terrigenous clastic depositional systems. New York: Springer-Verlag, 1983.

[32] Bend S L. Petroleum geology etextbook. 2008. Tulsa, OK, USA: AAPG.

[33] IEA (International Energy Agency). Resources to reserves 2013: oil, gas and coal technologies for the energy markets of the future. Paris, France: IEA. 155-196.

[34] Klett T R. United States Geological Survey's reserve-growth models and their implementation: Natural Resources Research, 2005, 14(3): 249-264.

[35] Magoon L B, Dow W G. The petroleum system // The petroleum system—from source to trap. AAPG Memoir 60, 1994: 3-24.

[36] Otis R M, Schneidermann N. A process for evaluating exploration prospects. AAPG Bulletin, 1997: 81(7), 1087-1109.

[37] Selley R C. Elements of petroleum geology . 3rd edition. London, UK: Academic Press, 1998.

[38] Sonnenberg S A, Mechkel L. Our current working model for unconventional tight petroleum systems: oil and gas. AAPG Pacific Section and Rocky Mountain Section Joint Meeting, Las Vegas, Nevada, 2016.

[39] Allen P A, Allen J R. Basin analysis: principles and applications. Oxford: Blackwell Scientific Publications, 1990.

[40] Zhang T F, Tilke P, Dupont E. Generating geologically realistic 3D reservoir facies models using deep learning of sedimentary architecture with generative adversarial

networks. Petroleum science, 2019, 16(3): 541–549.

[41] Hamawand I, Yusaf T, Hamawand S. Coal seam gas and associated water: A review paper. Renewable and Sustainable Energy Reviews, 2013, 22: 550–560.

[42] Kulga B, Artun E, Ertekin T. Characterization of tight-gas sand reservoirs from horizontal-well performance data using an inverse neural network. Journal of Natural Gas Science and Engineering, 2018, 59: 35–46.

Glossary

New Words

A(a)

abyssal *adj.* 深海的,深渊的
accredit *v.* 认可,信任,授权
accumulation *n.* 积聚,堆积物
acoustic *adj.* 听觉的,声波的
acreage *n.* 区块
adjacent *adj.* 临近的,接近的
adjoining *adj.* 毗连的,邻近的
aftershock *n.* 余震
aggregation *n.* 集合,集合体
algae *n.* 藻类, 海藻
alkaline *adj.* 碱的, 碱性的
alkane *n.* 烷烃
allocation *n.* 分配,配置
allochem *n.* 异化(颗)粒
alloy *n.* 合金
alluvium *n.* 冲积层
alternation *n.* 互层,交替
altitude *n.* (海拔)高度
aluminophosphate *n.* 铝磷酸盐
aluminum *n.* 铝
americium *n.* 镅
amino *adj.* 氨基的
ammonia *n.* 氨
ammonium *n.* 铵
amphibole *n.* 角闪石(类)
amplitude *n.* 振幅,波幅
ancestor *n.* 祖先,祖宗
anhydrite *n.* 硬石膏,无水石膏
anticline *n.* 背斜
antimony *n.* 锑
apatite *n.* 磷灰石
appendix *n.* 附录
applicable *adj.* 适用的,合适的
aprons *n.* 石裙
aqueous *adj.* 含水的,似水的
aquifer *n.* 蓄水层,含水层
aragonite *n.* 文石
archaeology *n.* 考古学
arenite *n.* 砂岩
argillite *n.* 泥岩,泥板岩
argument *n.* 证据,论据
arkoses *n.* 长石砂岩
aromatic *adj.* 芳香的
aromatization *n.* 芳烃化
arsenic *n.* 砷,砒霜
arsenide *n.* 砷化物
asphalt *n.* 沥青
assessment *n.* 评定,评价
assumption *n.* 假定,假设
asthenosphere *n.* 软流圈,岩流圈
astronomy *n.* 天文学
atlas *n.* 地图集,图解集
audit *vt.* 稽核, 审计
authigenic *adj.* 自生的
avalanche *n.* 雪崩
azimuth *n.* 方位角

B(b)

barite *n.* 重晶石
barium *n.* 钡
barrel *vt.* & *n.* 把……装入桶内,桶
basaltic *adj.* 玄武岩(质)的

basinal *n.* 盆地
bathyal *adj.* 半深海的
bauxite *n.* 铝矾土,铝土矿
beacon *n.* 信标,指向标
bedrock *n.* 岩基
beneficiation *n.* 选矿
benthic *adj.* 深海底的
beryllium *n.* 铍
bicarbonate *n.* 重碳酸盐
biodegrade *v.* 生物降解
biomass *n.* 生命体
biostratigraphy *n.* 生物地层学
biota *n.* 生物区
biotite *n.* 黑云母
bismuth *n.* 铋
blanket *n.* 毯子,覆盖层
borehole *n.* 井眼,钻眼
boron *n.* 硼
boulder *n.* 砾石,巨砾
breccia *n.* 角砾岩
brine *n.* 卤水,海水
brittle *adj.* 易碎的
bromine *n.* 溴
bury *vt.* 埋葬,隐藏

C(c)

calcarenite *n.* 砂屑灰岩
calcite *n.* 方解石
Cambrian *n.* & *adj.* 寒武纪(的)
canal *n.* & *vt.* 运河,开运河
capillary *adj.* 毛细管作用的
carbohydrate *n.* 碳水化合物
carbonate *n.* 碳酸盐
Carboniferous *n.* & *adj.* 石炭纪(的)
cataclastic *adj.* 压碎的,碎裂的
catagenesis *n.* 深成热解作用
catastrophism *n.* 灾难论
category *n.* 种类,类别,类型
caveat *n.* 警告
cement *n.* 胶结物
cementation *n.* 胶结作用
Cenozoic *n.* & *adj.* 新生代(的)
centrifuging *n.* 离心法,离心过滤
cesium *n.* 铯
cessation *n.* 中断,间断
chalcopyrite *n.* 黄铜矿
chaotic *adj.* 混乱的,紊乱的
characterization *n.* 描述
chenier *n.* 海岸沙脊
chert *n.* 燧石
chlorite *n.* 绿泥石
chronostratigraphy *n.* 地质年代学
cinder *n.* 火山渣,火山灰
cladistics *n.* 遗传分类学
clast *n.* 碎屑
clastic *adj.* 碎屑(状)的
clay *n.* 黏土
claystone *n.* 黏土岩
cleavage *n.* 解理,裂理
cobalt *n.* 钴
coefficient *n.* 系数
coevolution *n.* 共进化
collision *n.* 碰撞
colloidal *adj.* 胶质的,胶体的
colonize *vt.* 拓殖,殖民
commodity *n.* 日用品,商品
compaction *n.* 压实(作用)
compressional *adj.* 压缩的,受挤压的
compressor *n.* 压缩机
concentric *adj.* 同心的
concordant *adj.* 整合的,一致的
condensate *n.* 凝析油
conductivity *n.* 传导率
conglomerate *n.* 砾岩
consolidation *n.* 固结(作用)
contiguity *n.* 接近,邻近
contour *n.* 等高线,等值线
corroborative *adj.* 确定的

corundum　n. 刚玉,金刚砂
cosmochemistry　n. 宇宙化学
cosmology　n. 宇宙学
crack　n. 裂缝,缝隙,节理
Cretaceous　n. & adj. 白垩纪(的)
crossbedding　n. 交错层
crude　adj. 天然的,未加工的
crumple　vt. 变褶,弄褶
cryptocrystalline　adj. 隐晶质的
crystalline　adj. 结晶的
crystallization　n. 结晶作用
cubic　adj. 立方体的,立方的

D(d)

debris　n. 岩屑,碎屑
decipher　vt. 破译,解释
defoliation　n. 落叶
delimit　vt. 划定,界定
delineate　vt. 描写,描绘
delta　n. 三角洲
demarcate　v. 划分界限
deploy　v. 展开,部署
depocenter　n. 沉积中心
deposit　n. & vi. 沉积,矿床
depression　n. 洼地,凹陷
deterministic　adj. 决定的,确定的
detonate　vi. 起爆,爆炸
detrital　adj. 碎屑的
detritus　n. (侵蚀形成的)岩屑
devoid　adj. 全无的,缺乏的
Devonian　n. & adj. 泥盆纪(的)
diagenesis　n. 成岩作用
diamictites　n. 杂砾岩
diaper　n. 底辟(构造)
diapir　n. 底辟,挤入构造
diapirism　n. 底劈作用,刺穿作用
diffraction　n. 绕射,衍射
dike　n. 岩脉,堤坝,岩墙
dimension　n. 范围,尺寸,维度
dimensionality　n. 幅员,广延,维度
dinosaur　n. 恐龙
dioxide　n. 二氧化物
disclosure　n. 揭发,败露
disintegration　n. 崩解,解体
dislocation　n. 错位,变位
dispersed　adj. 分散的,散布的
dissipate　vt. & vi. 消散,消失
dissolution　n. 分散,溶解
dolomite　n. 白云岩
dolomitization　n. 白云石化作用
dormant　adj. 休眠的
dramatic　adj. 急剧的,戏剧性的
drilling　n. 钻探
dune　n. 沙丘
dyke　n. 岩墙

E(e)

eat　n. 泥炭,泥煤
echinoderm　n. 棘皮动物
echinoid　n. 海胆[类]
Ediacaran　n. 埃迪卡拉纪
electrode　n. 电极,电焊条
electrum　n. 银金矿,金银合金
emulsion　n. 悬浊液
encompass　vt. 包含,包围
endogenous　adj. 内长的,内生的
endowment　n. 储量,资源量
Eocene　n. & adj. 始新世(的)
eolian　adj. 风的
eon　n. 宙
epicenter　n. 震中
epigenesis　n. 晚期成岩作用
epoch　n. 新纪元, 世
equator　n. 赤道
equilibrium　n. 平衡,平静

era *n.* 时代,纪元,代
erosion *n.* 剥蚀,侵蚀
erratic *adj.* 不稳定的
estuarine *adj.* 河口的,江口的
estuary *n.* 河口,海湾
evaluate *vt.* 估定,评价
evaporite *n.* 蒸发岩(盐)
exinite *n.* 壳质组
exotic *adj.* 外来的
expansion *n.* 膨胀,扩张
exploration *n.* 勘探
exinite *n.* 壳质组
extraterrestrial *adj.* 地球外的,宇宙的

F(f)

facies *n.* 相,岩相
fairway *n.* 区带
fathometer *n.* 回音测深仪
fault *n.* & *vt.* 断层,产生断层
fauna *n.* 动物群
feature *n.* 特色,特征
feldspar *n.* 长石
fervent *adj.* 热情的,热心的
fissure *n.* 节理,裂缝,缝隙
fjord *n.* 海湾,峡湾
flexure *n.* 挠曲
flint *n.* 燧石,火石
flora *n.* 植物群
fluid *n.* 流体
fluorescence *n.* 荧光, 荧光性
fluvial *adj.* 河流的,河成的
fluxoturbidites *n.* 滑坡浊积岩
focus *n.* 焦点,焦距
fold *n.* 褶皱
forage *n.* 草料,饲料
foraminifera *n.* 有孔虫类
foredeep *n.* 前渊,山前洼地
formulate *vt.* 明确地表达
fossil *n.* & *adj.* 化石,化石的
foundation *n.* 基岩
fracture *n.* 裂缝
framboidal *adj.* 莓球状的
fungi *n.* 真菌类
fusibility *n.* 熔点

G(g)

galena *n.* 方铅矿
genesis *n.* 成因,起源
genome *n.* 基因组,染色体组
geochemistry *n.* 地球化学
geochronology *n.* 地质年代学
geology *n.* 地质学
geomechanical *adj.* 地质力学的
geometric *adj.* 几何(学)的
geomorphology *n.* 地貌学
geophone *n.* 检波器
geophysics *n.* 地球物理学
geostatistics *n.* 地质统计学
geotectonic *adj.* 大地构造的
geothermal *adj.* 地热的
glossary *n.* 词汇表,术语汇编
gneiss *n.* 片麻岩
graben *n.* 地堑
gradient *n.* 梯度
granite *n.* 花岗岩
granitoid *adj.* 花岗质的
granular *adj.* 粒状的
graphite *n.* 石墨
gravimetric *adj.* 重力测量的
gravitation *n.* 地心吸力
gravity *n.* 地心引力, 重力
greywacke *n.* 杂砂岩
gypsum *n.* 石膏,石膏肥料

H(h)

halides *n.* 岩盐,卤化物,石盐
harness *vt.* 利用,控制
helium *n.* 氦
herein *adv.* 在此处,如此
heterocyclic *adj.* 杂环的,不同环式的
heterogeneity *n.* 非均质性
heterogeneous *adj.* 异种的,非均质的
hiatus *n.* 裂隙,间断
Himalayas *n.* 喜马拉雅山脉
hinge *n.* (褶曲)枢纽
hinter *prep.* 在……面,到……面
homogeneous *adj.* 各向同性的
horst *n.* 地垒
humic *adj.* 腐殖的
humin *n.* 腐殖质,腐黑物
hydration *n.* 水合作用
hydraulic *adj.* 液压的
hydrocarbon *n.* 烃,碳氢化合物
hydrodynamic *adj.* 水动力学的,水力的
hydrogen *n.* 氢
hydrology *n.* 水文学
hydrophone *n.* 水中地震检波器
hydrosphere *n.* 水圈
hypothesis *n.* 假说,猜想
hypothetical *adj.* 可能的,假定的

I(i)

IES *abbr.* 声感测井
igneous *adj.* 火成的
ignimbrite *n.* 熔灰岩
illite *n.* 伊利石
immature *adj.* 未成熟的
immersion *n.* 沉浸
immiscible *adj.* 不相混的
impermeable *adj.* 不渗透的
impervious *adj.* 不可渗透的,透不过的
incandescent *adj.* 白炽的
incision *n.* 深切作用,下切作用
inclination *n.* 倾斜,倾角
inert *adj.* 惰性的
inherent *adj.* 内在的,固有的
initial *adj.* 最初的
insolation *n.* 日晒作用
insoluble *adj.* 不溶解的
interbedded *adj.* 夹层之间的
interconnection *n.* 互相联络
interlaminated *n.* 纹层状互层
interstellar *adj.* 星际的
intervene *vi.* 介于,插进
intricate *adj.* 复杂的
inundate *v.* 淹没
invertebrates *n.* 无脊椎动物
iodine *n.* 碘
irreducible *adj .* 束缚的,不能复归的
irreducibly *adv.* 不能减缩地,不能简化地
isochron *n.* 等时差线
isomorphism *n.* 类质同象
isopach *n.* 等厚线

J(j)

joint *n.* 节理,裂缝
jointly *adv.* 共同地,连带地
Jupiter *n.* 木星
Jurassic *n.* & *adj.* 侏罗纪(的)

K(k)

kaolinite *n.* 高岭石
karst *n.* 喀斯特,岩溶

Kazakhstan 哈萨克斯坦
kerogen *n.* 干酪根,油母岩
kyanite *n.* 蓝晶石

L(l)

laccolith *n.* 岩盘,岩盖
lacustrine *adj.* 湖泊的,湖相的
lagoon *n.* 潟湖
laminated *adj.* 纹层状的,叶片状的
lamination *n.* 层理
landsat *n.* 地球资源(探测)卫星
lapilli *n.* 火山砾
lateral *adj.* 侧面的,横向的
lava *n.* 熔岩
lead *n.* 铅
lenticular *adj.* 透镜状的,扁豆状的
lime *n.* 石灰,灰泥
limestone *n.* 石灰岩
lipid *n.* 脂质,类脂
liptinitic *adj.* 壳质组的
lithification *n.* 岩化作用,石化作用
lithologic *adj.* 岩性的
lithology *n.* 岩性
lithophagous *adj.* 食石的
lithosphere *n.* 岩石圈
lithostatic *adj.* 岩石静压力的
lithostratigraphy *n.* 岩性地层(学)
lobate *adj.* 舌状的,叶状的
log *n.* 测井,测井曲线
lunar *adj.* 月亮的,月球的
luidity *n.* 灰黄色
lutite *n.* 细屑岩

M(m)

magma *n.* 岩浆
magmatic *adj.* 岩浆的
magnesite *n.* 菱镁矿
magnesium *n.* 镁
magnetosphere *n.* 磁圈
manifestation *n.* 证明,显示
marcasite *n.* 白铁矿
marginal *adj.* 边缘的,临界的
marsh *n.* 沼泽,草沼
mass *n.* 块,团
mast *n.* 柱杆
matrix *n.* 基质
maturity *n.* 成熟,到期
mercury *n.* 水银, 汞
Mesozoic *n.* & *adj.* 中生代(的)
metagenesis *n.* 沉积变质作用
metamorphic *adj.* 变质的
metamorphism *n.* 变质作用
metasomatism *n.* 交代(作用)
meteoric *adj.* 大气的
meteorite *n.* 陨星,陨石
meteorology *n.* 气象学
methane *n.* 甲烷
micrite *n.* 泥晶,泥晶灰岩
microbial *adj.* 微生物的
migration *n.* 移动,迁移
mineral *n.* 矿物
mineralogical *adj.* 矿物学的
mineralogy *n.* 矿物学
miocene *n.* & *adj.* 中新世(的)
Moho *abbr.* 莫霍面
molecular *adj.* 分子的
molefraction *n.* 克分子比
mollusk *n.* 软体动物
monocline *n.* 单斜(层)
monocrystalline *n.* 单晶(体)
morphology *n.* 形态学,形态论
motif *n.* 主题,图形
mud *n.* 泥,泥浆
mudslide *n.* 泥石流

multidisciplinary *adj.* 多学科的

muscovite *n.* 白云母

N(n)

naphthenic *adj.* 环烷的

nappe *n.* 推覆体,逆掩褶曲

Neogene *n.* &*adj.* 新近纪(的)

neomorphism *n.* 新生变形作用

Neptune *n.* 海王星

neritic *adj.* 浅海的,近岸的

nitrogen *n.* 氮

nonrenewable *adj.* 不可再生的

O(o)

oceanography *n.* 海洋学

OECD *abbr.* 经济合作和发展组织

offshore *adj.* 近海的

oilfield *n.* 油田

olivine *n.* 橄榄石

oncoid *n.* 似核形石

oolite *n.* 鲕粒石灰岩

oolith *n.* 鲕粒

OPEC *abbr.* 欧佩克

ophiolite *n.* 蛇绿岩,蛇绿岩系

Ordovician *n.* & *adj.* 奥陶纪(的)

ore *n.* 矿,矿石

organic *adj.* 有机的

organism *n.* 生物体,有机体

orogenic *adj.* 造山的,造山作用的

orogeny *n.* 造山运动

orthoquartzite *n.* 正石英砂岩

overlap *v.* (与……)交搭,部分一致

overthrust *n.* 逆掩断层,上冲断层

oxidation *n.* 氧化(作用)

ozokerite *n.* 地蜡,石蜡

P(p)

palaeocurrent *n.* 古水流

paleocanyons *n.* 古峡谷

Paleocene *n.* & *adj.* 古新世(的)

paleocliff *n.* 海崖

paleoecology *n.* 古生态学

Paleogene *n.* & *adj.* 古近纪(的)

paleogeography *n.* 古地理学

paleomagnetism *n.* 古地磁(学)

paleontologic *adj.* 古生物的

paleontological *adj.* 古生物(学)的

paleontology *n.* 古生物学

paleotopography *n.* 古地形

paleovalley *n.* 古谷

Paleozoic *n.* &*adj.* 古生代(的)

paraffin *n.* 石蜡

parametric *adj.* 参数的,变量的

parasequence *n.* 副层序,准层序

peloid *n.* 球粒

penecontemporaneous *adj.* 准同生的

peneplain *n.* 准平原

peneplane *n.* 准平原

penetrate *vt.* 穿透

perceive *vt.* 理解,认知

percolating *n.* 渗透

perforation *n.* 空隙

perimeter *n.* 周长,周边

permafrost *n.* 永久冻结带

permeability *n.* 渗透性,渗透率

Permian *n.* & *adj.* 二叠纪(的)

petrographic *adj.* 岩相学的,岩相的

petrography *n.* 岩相学

petrology *n.* 岩石学

petrophysics *n.* 岩石物理学

Phanerozoic *adj.* 显生宙的

phosphate *n.* 磷酸盐

phosphating *n.* 磷酸盐处理

phyllosilicate *n.* 页硅酸盐
phylogenetics *n.* 系统发生学
physicochemical *adj.* 物理化学的
phytoplankton *n.* 浮游植物
pisoid *n.* 豆粒
plan *n.* 设计图,平面图
platinum *n* 白金, 铂
playa *n.* 干盐湖,干荒盆地
Pliocene *n.* & *adj.* 上新世(的)
plugged *adj.* 被堵塞的
pollinate *vt.* 对……授粉
polymer *n.* 聚合体
polymeric *adj.* 聚合的,聚合体的
porosity *n.* 多孔性,有孔性
porous *adj.* 多孔的
portfolio *n.* 组合
postdepositional *adj.* 沉积后的
postulate *vt .* 要求,假定
potassium *n.* 钾
Precambrian *n.* & *adj.* 前寒武纪(的)
precipice *n.* 悬崖,峭壁
precipitate *vt.* 沉淀,结晶
precipitation *n.* 降雨的量
precursor *n.* 先驱,先兆
prehistoric *adj.* 史前的,陈旧的
preservation *n.* 保存,保护
principal *adj.* 首要的,主要的
pristine *n.* 姥鲛烷
probabilistic *adj.* 概率的,可能性的
progressively *adv.* 日益增多地
projection *n.* 投影,投射
prominent *adj.* 突出的,显著的
propane *n.* 丙烷
property *n.* 特性,性能
proton *n.* 质子
protuberance *n.* 突起
provision *n.* 供应
pyrite *n.* 黄铁矿
pyroclastic *adj.* 火山碎屑的
pyroxene *n.* 辉石

Q(q)

quantitatively *adv.* 定量地,数量上
Quaternary *n.* & *adj.* 第四纪(的)

R(r)

radioactive *adj.* 放射性的
radiometric *adj.* 放射测量的
real-time *n.* & *adj.* 实时(的)
reciprocal *adj.* 互惠的,互补的
reclamation *n.* 回收
recombined *adj.* 调配的
recovery *n.* 采收率, 恢复
recycling *n.* 重复使用,回收
reef *n.* 生物礁
reflectance *n.* 反射率
refractive *adj.* 折射的
regime *n.* 沉积环境
regression *n.* 海退,退化
release *n.* 发表,释放
relief *n.* 地形,地貌
renaissance *n.* 文艺复兴(时期)
renewable *adj.* 可再生的
repetition *n.* 重复
reptile *n.* 爬虫动物
reservoir *n.* 储层,油藏
resiny *adj.* 树脂质的,含树脂的
rhombohedral *adj.* 斜方六面体的
rift *n.* 裂缝,裂隙
rigidity *n.* 刚度
rigorous *adj.* 严格的,严厉的
ripple *n.* 波痕
robustness *n.* 坚实,强度
rotary *adj.* 旋转的,轮流的

rotate vi. 旋转
rudite n. 砾屑岩
rugosity n. 不规则,皱褶
rupture n. 断裂,破裂,节理
rutile n. 金红石

S(s)

sabkha n. 盐滩,萨布哈,潮上滩
salinity n. 盐分,盐度
salt n. 盐
saltation n. 跃移
saturate vt. 饱和,浸透
saturation n. 饱和(状态),饱和度
scenario n. 成果,方案
schist n. 片岩,结晶片岩
seal n. 盖层
sealed adj. 封口的,密封的
section n. (地层)剖面
sediment n. 沉积物
sedimentary adj. 沉积的
sedimentation n. 沉积
sedimentological adj. 沉积学的
sedimentology n. 沉积学
seep n. & v. 油气苗,渗漏
seismic adj. 地震的
seismics n. 地震学
seismograph n. 地震仪
seismologist n. 地震学家
seismology n. 地震学
sequence n. 层序,序列
sewage n. 污水,垃圾
shale n. 页岩
shallow adj. 浅的
shield n. 盾,防护物
sideswipe vt. 擦边撞击,沿边擦过
silica n. 二氧化硅
silicate n. 硅酸盐
siliciclastic adj. 硅质碎屑的
silicification n. 硅化(作用)
silt n. 粉砂
siltstone n. 粉砂岩
sima n. 硅镁层
skeletal adj. 骨骼的
slope n. 斜坡,倾角
slump n. 滑塌
solidification n. 凝固,固化
solifluction n. 泥流
solubility n. 溶解度
soluble adj. 可溶的,可溶解的
sonde n. 探测器,探头
sophisticated adj. 复杂的,尖端的
sorb vt. 吸附,吸收
SP abbr. 自然电位
span n. 跨度,范围
sparite n. 亮晶,亮晶方解石
speciation n. 物种形成
spectrum n. 范围;系列
speculative adj. 推测的,推理的
sphalerit n. 闪锌矿
squeeze vt. 挤,压榨
stacked adj. 叠加的
stagnant adj. 停滞的
strand n. 海滨,潮间带
strandplain n. 海滨平原
strata n. 地层
stratification n. 成层,层理
stratigrapher n. 地层学家
stratigraphy n. 地层学
stromatolite n. 叠层(石)
strontium n. 锶
structural adj. 结构的,建筑的
stylolitization n. 新生变形作用
subaerial adj. 地面上的
subaqueous adj. 水下的
subarkose n. 亚长石砂岩
subcrustal adj. 地壳之下的
subdiscipline n. 分支学科

subdivide *vt.* 再分,细分
subjective *adj.* 主观的,主观上的
sublitharenite *n.* 次岩屑砂岩
subsequent *adj.* 后来的,随后的
subsidence *n.* 下沉,沉降
substrate *n.* 基质,底层
subterranean *adj.* 地下的
subtle *adj.* 不可思议的,微细的
suite *n.* 岩系
sulfate *n.* 硫酸盐
sulfide *n.* 硫化物
superimpose *vt.* 使重叠,使叠加
supersaturate *vt.* 使过饱和
surfactant *n.* 表面活性剂
suspension *n.* 悬浮液
swivel *n.* 转体
sylvite *n.* 钾盐
syndepositional *adj.* 同沉积的
synergy *n.* 协同作用
synonym *n.* 同义词,同物异名

T(t)

tabulate *vt.* 把……制成表格
tectonism *n.* 构造作用
telocollinite *n.* 均质镜质体
tension *n.* 应力状态
term *n.* 期限,术语,条款
terrestrial *adj.* 陆地的,陆相的
terrigenous *adj.* 陆源的,陆生的
Tertiary *n.* 古近—新近纪
texture *n.* 结构
the magnetism *n.* 磁性,磁学
theoretical *adj.* 理论的
thrive *v.* 兴旺,繁荣
thrust *n.* & *vt.* 挤入,逆冲
thumper *n.* 重击声,雷声
tidalite *n.* 潮积岩
till *n.* 冰碛物
topography *n.* 地形,地形学
tow *v.* 拖,拽
traction *n.* 牵引作用
transgression *n.* 海进, 海侵
transparency *n.* 透明度
trap *n.* 圈闭,陷阱
trench *n.* 海沟, 深沟
tricky *adj.* 复杂的,棘手的
trigger *vt.* 触发,发射
trough *n.* 槽,沟
truncate *vt.* 切去头端,截去……顶端
tsunami *n.* 海啸
turbidite *n.* 浊流岩
turbidity *n.* 浊流

U(u)

unconformity *n.* 不整合,不整合面
undercutting *n.* 切挖,钻蚀
underlying *adj.* 下伏的,潜在的
unidirectional *adj.* 单向的
uniformitarianism *n.* 均变论
upset *vt.* 翻倒,颠覆
upthrust *n.* 逆断层,(上)冲断层
upward *adj.* 向上的,上升的

V(v)

Vail curve *n.* 凡尔(海平面)曲线
variegated *adj.* 杂色的,多样化的
veneer *n.* 薄板,外表
Venus *n.* 金星
verification *n.* 确认,作证,查证
vertical *adj.* 垂直的
viability *n.* 总产量,总收率
viable *adj.* 可行的,能生育的
viscosity *n.* 黏度
viscous *adj.* 黏性的

vitrinite *n.* 镜质组,镜煤组
volcano *n.* 火山
volume *n.* 体积
volumetric *adj.* 体积的,容积的
vug *n.* 孔洞

W(w)

wadi *n.* 旱谷,干谷
weathering *n.* 风化(作用),侵蚀
wellbore *n.* 钻井孔,打井孔
wettability *n.* 润湿性
wiggle *n.* 波跳动
winnow *vt.* 扬,簸,筛选

Z(z)

zircon *n.* 锆石
zooplankton *n.* 浮游动物

Phrases

A(a)

a demolition derby 一场破坏性比赛
a disruption in sedimentation 沉积间断
a flowing salt glacier 流动的盐冰川
a hard and fast 一板一眼的,严厉的
a malleable layer 可塑层
a plate boundary 板块边缘(界)
a silicated viscous matrix 黏性硅酸盐基质
abruptly with vitrification 突然玻璃化
accommodation space 可容空间
accumulation-and-migration 成藏和运移
acoustic property 声学特征
active rift 活动断裂,活动裂谷
adsorbed gas 吸附气
aeolian environment 风成环境
alkali-earth 碱土金属
alluvial fan 冲积扇
alternative source 替代资源
anaerobic digestion process 厌氧消化工艺
anaerobic microorganisms 厌氧微生物
an expansive advance of the sea 海进
an overall subsidence of land 陆地整体沉降
ancient tectonic movement 古代构造运动
Andes Mountains 安第斯山脉
angular unconformity 角度不整合
anticlinal fold 背斜形褶曲
arenite-sand 净砂岩砂粒
argillaceous silt 泥质粉砂(岩)
arkosic arenite 长石砂屑岩
arkosic sandstone 长石质砂岩
artificial neural network 人工神经网络
artificial stimulation 人工改造
as opposed to 相对于
asymmetric anticline 不对称背斜
asymmetrical fold 不对称褶曲
at a right angle 成直角

B(b)

barrier bar 障壁坝
barrier island 屏障岛
basaltic lava 玄武质熔岩
be perpendicular to 垂直于
bed load 底负载
bedding plane 层面
bedding plane 层面,层理面
bedding surface 层面,层理面

beyond the scope 超出范围
biogenic gas 生物气
biomarker analysis 生物标志物分析
black shale formations 黑色页岩层
blind joint 隐蔽节理
blunt fold 钝状褶曲
borehole data 钻井数据
botryoidal grain 葡萄状颗粒
bottom water 底水
bow tie effect 蝴蝶结效应
bowl-like fold 似碗状褶曲
brachy-formed fold 短轴褶曲
braid bar 网状河沙坝
braided river 网状河
break fissure 张性节理(裂理)
breakdown pressure 临界压力
brittle minerals 脆性矿物
building operations 建筑施工
burial history 埋藏史

C(c)

calcareous mudstone 钙质泥岩,石灰质泥岩
calcareous sandstone 钙质砂岩
calcium carbonate 碳酸钙
calcium magnesium carbonate 碳酸钙镁
carbonate ramp 碳酸盐斜坡,碳酸盐坡地
carbonate reefs 碳酸盐礁
Carboniferous limestone 石炭纪灰岩
Carboniferous period 石炭纪
cardinal points 基本方位
Caspian area 里海地区
catchment area 汇水面积,集水区
caustic soda 苛性钠
celestial body 天体
centroclinal fold 向心褶皱
centroclinal structure 同心向斜构造
channel bar 河道沙坝,心滩
channel fill 凹槽充填,水道充填
chemical degradation 化学降解
chemical flood 化学驱油
chert clasts 硅质碎屑
chlorine clathrate 氯包合物
chronostratigraphic unit 年代地层单位
circulating fluid 循环液
claim for payment 要求付款
clastic coasts 碎屑海岸
clastic reservoir 碎屑岩储层
clay mineral reaction 黏土矿物作用
clay mineral 黏土矿物
coalbed methane 煤层气
coal-seam gas 煤层气
coffer (box) fold 箱状褶曲
columnar joint 柱状节理
combination trap 复合圈闭
common depth point 共深点
comparative anatomy 比较解剖学
complementary tool 补充工具
compressional stress 压应力
concentric fold 同心状褶曲
condensed sequence 凝缩层序
conformable succession 整合序列
connate water 原生水
Conrad discontinuity 康拉德界面
considerable impact 巨大影响
considerably less often 相当少,很少
consolidated geological maps 综合地质图
contemporaneous tectonic 现代构造
continental drift 大陆漂移
continental margin 大陆边缘
continental plate 大陆板块
continental sedimentary series 大陆沉积岩系
contour map 等值线图
convection currents 对流
conventional natural gas 常规天然气
conventional reservoirs 常规油气藏
convergent plate boundary 会聚板块边缘
core analysis 岩心分析

core box　岩心箱,岩心盒
core lifter　岩心提取器,取心器
coupled with　连同,与……一起
created for this purpose　为此目的创建
crevasse splay deposition　决口扇沉积
critical moment　关键时刻
cross section　横剖面
crude oil　原油
crustal plate　地壳板块
cubic joint　立方节理
current-scoured shelf　水流冲蚀大陆架

D(d)

data-driven　数据驱动
debris flow　泥石流,碎屑流
decline-curve coefficient　衰减曲线系数
deep sea trenches　深部海沟
deep water　深水
deeper burial　埋藏深的
deltaic distributary channel　三角洲分流河道
density contrast　密度对比,密度差异
denudation and sedimentation　剥蚀和沉积作用
depletion drive　溶解气驱
depositional environment　沉积环境
derive from　来源于,来自
derrick board　井架板
detailed geological maps　详细地质图
development geology　开发地质学
diagenetic factor　成岩因素
diesel fuel　柴油
dissolved gas　溶解气
different perspectives　不同的角度
direct /indirect indicator　直接/间接标志
dome-shaped fold　穹窿构造
drawdown test　压降测试
drill bit　钻头
drill collar　钻铤
drill crew　钻井队,钻探班组
drilling mud　钻井液
drilling rigs　钻机
driving force　驱动力
driving mechanism　驱动机制
dry well　干井,枯井
due in part to　部分由于,归因于
dynamic viscosity　动力黏度
dynamo theory　(地磁场成因的)发电机理论

E(e)

earthquake hazard　地震灾害
Earth's hot interior　热的地球内部
eastern Carpathians　(中欧)东喀尔巴阡山脉
economic geology　经济地质学(矿床学)
edge water　边水
elastic properties　弹性
electricity generation　发电量
elemental abundance　元素丰度
elements of bedding　岩层(产状)要素
empirical equation　经验公式
endogenic processes　内营力作用
enhanced oil recovery　提高原油采收率
environments of deposition　沉积环境
ephemeral lake　季节湖
equal-chance game　等概率游戏
evaporite deposit　蒸发岩沉积
evaporite minerals　蒸发岩矿物
even with　即使
exogenic processes　外营力作用
explanatory note　注释
exploration geophysicist　勘探地球物理学家
expulsion force　排驱力
extinction event　(生物)灭绝事件
extra-heavy oil reserves　超重质油储量
extrusive rock　喷出岩

F(f)

fairway for the play　成藏组合区带
fall into　分成,变成,落入
fan-shaped fold　扇形褶曲
fault blocks　断块(构造)
fault zones　断裂带
fault-related energy　断层相关的能源
faunal succession　动物区系演替
ferromagnetic crystals　铁磁晶体
ferromagnetic minerals　铁磁矿物
fissure of weathering　风华节理
fissures and ruptures　节理与断层
flat joint　平节理
floodplain deposit　泛滥平原沉积
fluid saturation　流体饱和度
fold of the oval shape　椭圆形褶曲
folded trough　褶皱凹陷
formation fluid　地层流体
formation pressure　地层压力
fossil fuel　化石燃料
fossil compasses　化石罗盘
fossil land vertebrates　陆地脊椎动物化石
free gas　游离气
freeze-thaw　冻融
frontier basin　边际盆地
functions of geological map　地质图的用途

G(g)

gaping fault　张开断层
gaps and overlaps　空白带和重叠区
gas bubble　气泡
gas cap drive　气顶气驱动
gas hydrate　天然气水合物
gas injection　注气
gas-oil ratio　气油比
genesis of fissure　节理成因
generation and migration of hydrocarbons　烃类的生成与运移
genetic character　成因特征
genetic increment　成因增量
genetic relationship　成因联系
gently sloping upthrust　平缓逆断层
geochemical match　地球化学可比性
geochronologic unit　地质年代单位
geodetic measurement　大地测量
geographical unconformity　地理不整合
geologic characteristic　地质特征
geological and engineering maps　工程地质图
geological section　地质剖面
geological variation　地质变化
geomagnetic anomalies　地磁异常
geomagnetic pole　地磁极(点)
geomorphological maps　地貌图
geophysical well logs　地球物理测井
geostatic pressure　地层压力
geostatistical methods　地质统计学方法
geotectonic maps　大地构造图
geothermal energy　地热能
geothermal gradient　地温梯度
geothermal reservoir　地热储层
geothermal resources　地热资源
glacial deposit　冰川沉积
glacial material　冰碛物
glass shard　玻屑
gliding plane　滑动面
globular joint　球状节理
grain size　粒度, 颗粒尺寸
gravitational forces　重力
gravitational phenomena　重力现象
gravitational slide phenomena　重力滑动现象
gravity drainage drive　重力排水驱动
Greenland Shelf　格陵兰大陆架
Greenland　格陵兰岛
ground masses　地面岩体

growth faults　生长断层

H(h)

heat conductivity　热传导率
heat-injection　热注入
heavy oils　重质油
heterogeneous masses of rock　非均质岩体
high recovery　高采收率
high stand tract　高水位体系域
hinge of fold　褶曲枢纽
history-matching process　历史拟合过程
homogeneous rock　均质岩石
horizontal drilling　水平钻井
horizontal joint　水平节理
horizontal movement　水平运动
horizontal or multilateral wells　水平井或者丛式井
the flat pitch of layers　水平岩层
hot volatile rich magma　富含挥发分热岩浆
huff and puff　蒸汽吞吐
hydraulic fracturing　水力压裂
hydraulic-fracture　水力压裂
hydraulic gradient　水力梯度
hydraulic pump　液压泵
hydrocarbon accumulation　油气藏(田)
hydrocarbon potential　油气潜力
hydrogen sulfide　硫化氢
hydrogeological maps　水文地质图
hydrostatic pressure　静水压力
hydrothermal vent　深海热泉

I(i)

idealized mushroom-shaped　理想化的蘑菇状
igneous intrusion　岩浆的侵入
inert gases　惰性气体
impermeable rock　不渗透岩层
in accordance with　与……一致,依照
in conjunction with　连同,共同
in question　正在考虑的
in this context　在这种背景下
inclined fold　倾斜褶曲
indication of discontinuity　无缺失显示
indiscriminately averaging　大平均
indurated shales　硬结页岩
injecting geothermal resource　注入地热资源
in-situ combustion　火烧油层
intergranular pores　晶粒间孔隙
international nomenclature　国际命名法
interstitial water　隙间水
interval transit time　穿层时间
intrabedded deformation　交互层间变形
intracratonic basin　克拉通内盆地
intramontane basin　山间盆地
intrusive rock　侵入岩
inverted petroleum system　反转含油气系统
iron oxide　氧化铁
irrefutably prove　无可辨驳地证明
island arcs　岛弧
isoclinal fold　等斜褶曲
isomorphous series　类质同象系列
isopach map　等厚图

K(k)

kandite group　高岭石组
keen demand　强烈的需求
kerogen shale　干酪根页岩,油页岩
kelly bushing　方钻杆补心
kinds of geological maps　地质图的种类
kinematic evolution 动态演化过程

L(l)

lamination surface 层面
landslide and sink fissures 滑塌或沉降节理
large bulbous dome 大型球形穹隆
lateral accretion 侧向增生,侧向加积(作用)
lateral continuity 侧向连续性
lateral direction 横向
lateral permeability 横向渗透率
law of superposition 层序律
liberate alkali 解放碱
light tight oil 轻质致密油
linear extrapolation 线性外推
linear fold 线状褶曲
linear schistosity 线状片理
linear valleys 线状山谷
liquefaction-fluidization 液化流
lithic arenite 砂屑岩
lithic fragment 石屑
lithologic characteristics 岩石学特性
lithological maps 岩相(性)图
living matter 生物质
lognormal distribution 正态分布
longitudinal fissures 纵向节理
Lorenz coefficient 洛伦茨系数
low permeability 低渗透率
low stand tract 低水位体系域
low tectonic stress 低构造应力

M(m)

magmatic intrusions 岩浆侵入体
magmatic processes 岩浆作用
maps of mineral resources 矿产资源图
marine carbonaceous silt 海相碳质泥沙
marine sedimentary rocks 海相沉积岩
marine shales 海相页岩
marker horizon 标志层
massive composition 大块岩体(岩基)
maximum flooding surface 最大海泛面
meandering river deposit 曲流河沉积
mechanical behavior 力学特性,机械特性
Mercalli scale 麦加利震级
metamorphic rock 变质岩
methane hydrates 甲烷水合物
microbial EOR 微生物强化采油
microseismic survey 微地震测量
microwave and radio frequency 微波和射频
mid-oceanic ridges 大洋中脊
migration avenues 运移通道
mineral reserves 矿产资源
miscible hydrocarbon displacement 烃类混相驱
miscible recovery processes 混相采收工艺
misleading interpretation 误导性解释
Mohorovicic discontinuity 莫霍不连续面
monoclinal bedding 单斜岩层
Monte Carlo simulation 蒙特卡罗模拟
morphological arguments 形态证据
mountain ranges 山脉
mud diapir 泥底辟
mud pump 钻井泵
multi-layered continental 陆相多油层的

N(n)

natural gas-hydrates 天然气水合物
Navajo sandstone 沙丘砂岩
non-tectonic fissures 非构造节理
non-tectonic genesis 非构造成因
normal fault 正断层
normal fold 正常褶曲
nuee ardentes 白炽灯云

O(o)

oblique fissures　斜节理
ocean trench　海沟
oceanic crust　大洋地壳,海洋地壳
oceanic plate　大洋板块
oceanic topography　海洋地形
oi sand body　油砂体
oil equivalent　石油当量
oil recovery processes　采油工艺
oil recovery　采油
oil sands　油砂
oil shale　油页岩
oil window　生油窗
oil-based mud　油基钻井液
open fractures　开放性裂缝
open joint　张节理
ores and mineral springs　矿床和矿脉
organic fraction　有机馏分
organic matters　有机质
organic maturity　有机质成熟度
organic reaction　有机作用
original fissure　原生节理
orogenic rock　造山带岩石
orthorhombic crystal system　斜方晶系
overburden rock　上覆岩层
overlying layer　上覆地层
overpressured compartment　超压箱
overpressuredfloater　超压流体
overthrusts and nappes　逆掩断层
overturned fold　倒转褶曲
overturned folds　倒转褶皱

P(p)

palaeomagnetic field　古地磁场
parallel unconformity　平行不整合
particular instance　特例
pay zone　产油层
pebbly mudstone　石蛋泥石岩
pelagic detritus　海面碎屑
perforated casing　射孔的套管
performance characteristic　性能特征
performance measure　绩效衡量
permeability variation　渗透率变异
petroleum column　油柱
petroleum engineering　石油工程
petroleum entrapment　石油圈闭
petroleum industry　石油工业
petro-physics　油层物理
photic zone　透光层
piercement structure　刺穿构造
pilot borehole　试验钻孔
plagioclase feldspar　斜长石
planar cross-bedded　板状交错层理的
plane schistosity　面状片理
plate boundary　板块边界
plate collision　板块碰撞
plate margin　板块边缘
plate separation　板块分裂(离)
plate tectonics　板块学说
plate tectonics　板块构造
platy joint　板状节理
polymer flooding　聚合物驱油
pore space　孔隙,孔隙空间
porous medium　多孔介质
porous rock　多孔岩石
possible reserve　预测储量
potassium feldspar　钾长石
potassium iodate　碘酸钾
precipitation of gypsum　石膏沉淀析出
pressure buildup test　压力恢复测试
pressure solution　压溶
primary migration　初次运移
primary oil recovery　初次采油
primary porosity　原生孔隙度
primary reflector　一次反射层

prismatic joint 棱柱状的节理
probable reserve 控制储量
produced-water reinjection 污水回注
producing formation （生）产层
prognostic maps 预测图
prograding delta 前积三角洲
prospecting and exploration 普查与勘探
prospective shale gas basins 有远景的页岩气盆地
proved reserve 探明储量
purifying the gas 净化气体

R(r)

radioactive isotope 放射性同位素
rate of subsidence 沉陷速度
recent tectonic movement 近代构造运动
recording truck 仪器车,记录车
recoverable bitumen reserves 可采沥青储量
recoverable reserves 可采储量
rectilinear joint 线状节理
recumbent fold 平卧褶曲
reflecting horizon 反射层
regional plate tectonic event 局部板块构造活动
remote sensing 遥感
renewable energy 可再生能源
reserve growth 拓展储量
reservoir behavior 储层动态
reservoir characterization 储层表征
reservoir description 油藏描述
reservoir pressures 储层压力
reservoir properties 储层属性
reservoir rock 储层
resource-demanding 资源需求
residual carbon 残余碳
residual oil 残余油,剩余油
restricted marine basins 封闭的洋盆
returns 返出液
reverse fault 逆断层
Richter scale 里氏震级
rig deck 钻机甲板
rigid pieces 刚性板块
rip current 裂流,离岸流
roll-over anticline 滚动背斜

S(s)

salt dome 盐丘,盐穹,岩丘
salt plug 盐栓(盐丘)
schistose rocks 片状岩石
screening device 筛选方法,筛选手段
seafloor spreading 海底扩张
secondary migration 二次运移
secondary porosity 次生孔隙
secondary recovery 二次采油
secondary replacement 次生交代
sedimentary cover 沉积盖层
sedimentary fill 沉积充填
sedimentary rock column 沉积岩剖面
sedimentary rock 沉积岩
sedimentation rate 沉积速率
seismic reflection profiling 地震反射剖面
seismic survey 地震勘探
seismic waves 地震波
sequence boundary 层序界面,层序边界
sequence stratigraphy 层序地层学
shale gas 页岩气
shallow burial 埋藏浅的
shallow invasion 浅层侵入
sharp fold 尖棱褶曲
shear joint 剪切节理
shear fault 剪切断层
shearing stress 抗剪应力
sheet flood 片流
shoreface facies 滨面相
shot hole 炮眼,炮井
shutin pressure 关井压力

siliciclastic coastlines　碎屑的海岸线
similar fold　相似形褶曲
size grading　粒径分级
sketch or areal maps　区域地质图
slab joint　板状节理
slaty cleavage　板岩劈理
slow decay of radioactive particles　反身性粒子的缓慢衰变
small-volume magmas　小批量岩浆
smectite group　蒙脱石组
solar system　太阳系
some mantle plumes　一些地幔柱
source rock　油源岩,母岩
space probe　空间探测器
space shuttle　航天飞机
spatial heterogeneity　空间非均质性
spatial resolution　空间分辨率
special nomogram　特别诺模图
spill point　溢出点
spin-axe　旋转轴
steam flooding　蒸汽驱
steep joint　陡峭节理
steep upthrust　陡峭逆断层
stochastic modeling　随机模型
storage geometry　储集空间
stratigraphic break　地层间断
stratigraphic column　地层柱状图
stratigraphic horizon　地层层位
stratigraphic record　地层记录
stratigraphic section　地层剖面
stratigraphic succession　地层层序
stratigraphic trap　地层圈闭
stratigraphic unconformity　地层不整合
stratigraphic unit　地层单位
stream bed　河床
strike joint　走向节理
structural arguments　构造证据
structural geology　构造地质学
structural trap　构造圈闭
structure contour map　构造等高线图
subaqueous condition　水下条件
subduction zone　俯冲带
subglacial channels　冰川渠道
submarine canyon　海底峡谷
subsurface data　地下数据
submarine fans　海底扇
superimposed layers　叠加岩层
supratenuous fold　顶薄形褶曲
surface tension　表面张力
surrounding rock　围岩
supplementary enhancement method　补充增强方法
sweet spots　甜点
swelling clay　膨胀黏土
symmetrical folds　对称褶曲
synclinal fold　向斜形褶曲
systems tract　体系域

T(t)

technically recoverable shale gas reserves　技术可采页岩气储量
tectonic breccia　构造角砾岩
tectonic contact　构造面(界限)
tectonic disturbance　构造变动,构造扰动
tectonic environment　构造环境
tectonic maps　构造图
tectonic movement　构造运动
tectonic stress field　构造应力场
tectonic unconformity　构造不整合
tensional stress　张应力
tertiary oil recovery　三次采油
tertiary techniques　三采技术
the abyssal mantle　深部地幔
the amplitude of displacement　断距
the ancient suite　古老岩层
the angle of inclination　倾角
the angle of unconformity　不整合角度

the Appalachian area　阿巴拉契亚地区
the apparent thickness　视厚度
the axial line　（褶曲）轴线
the axial plane　（褶曲）轴面
the azimuth of dip　倾向方位
the azimuth of strike　走向方位
the basal conglomerate　底砾岩
the basaltic layer　玄武岩层
the beaten path　因循守旧
the cessation of sedimentation　沉积间断
the composition of layered strata　层状岩层
the composition of rock　岩石组合
the contact plane　（不整合）接触面
the continental condition　大陆条件（环境）
the continental Earth's crust　大陆地壳
the continental masses　大陆块
the contour line　等高线
the core of the fold　褶曲的核部
the cross dimensions　纵横维度
the disharmonic suite　奇形怪状岩层
the dislocation rock　岩石的变形（位）
the displacements of magma　岩浆的移动
the Earth's magnetic field　地球磁场
ancient platforms　古地台
the field register　野外记录本
the fold axis　褶曲的轴
the genetic classification　成因分类
the geological medium　地质介质
the geosynclinal region　地槽区
the Gulf of Mexico　墨西哥湾
the Hercynian range　海西期山脉
the hinges of folds　褶皱的枢纽
the hydrate saturated layers　天然气水合物饱和层
the intrusive massifs　侵入体
the laminated strata　成层状岩层
the lamination of the suite　（一套）岩层
the lamination surface　岩层层面
the leading fossils　指相化石
the line of dip　倾向线
the line of strike　走向线
the lithological composition　岩性组合
the long axis of fold　褶曲的长轴
the lower suite　下部岩层
the Mid Atlantic ridge　中大西洋海岭
the morphological classification　形态分类
the mountain crest　山脊
the oceanic Nazca Plate　纳斯卡大洋板块
the oceanic slopes　海洋斜坡
the overlapping layers　上覆岩层
the overlying granitic layer　上覆花岗岩质层
the overlying suite　上覆岩层
the periclinal closure　穹隆构造
the position of axial surface　轴面产状
the process of diagenesis　成岩作用
the process of folding　褶皱过程
the process of sedimentation　沉积作用
the roof and the sole of a layer　岩层的顶底
the sedimentary sequences　沉积层序
the stratigraphic subdivision　地层
the stratigraphic sequence　地层层序
the stratigraphic subsection　部分地层
the strike of the fault fissure　断层面的走向
the strike of the layer　岩层走向
the supraasthenospheric layer　上软流圈层
the surface of disruption　断裂面
the surface of fault　断层面
subaqueous washout　水下冲刷
terrestrial denudation　地表剥蚀
the surface of unconformity　不整合面
the tectonic dislocation　构造变位
the time of elevation　（地壳）抬升期
the transgressive bedding　海侵地层
the true thickness of a layer　岩层真厚度
the underlying layers　下伏岩层
the undulated bend　波状弯曲
the upper suite　上部岩层
the U. S. Energy Information Administration (EIA)　美国能源信息管理局
the U. S. Geological Survey　美国地质调查局
the walls of fold　褶皱的翼部

the outcrop of the layer　岩层露头
the young suite　年轻地层
the Zagros Mountains　扎格罗斯山脉(伊朗)
thermal contraction 热收缩
thermal recovery processes　热力采油
thermodynamic conditions　热力学条件
three vertical boreholes　三口垂直钻孔
tidal flat　潮坪,潮滩
tidal inlets　潮口
tight fracture　压性节理
tight-gas　致密气
tight reservoir　致密气藏
inclined (or monoclinal) bedding　倾斜岩层
tooth tri-cone　齿牙轮
topographic map　地形图
transgressive tract　海侵体系域
transpressional thrust belt　走滑挤压逆冲带
transverse fissures　横向节理(裂缝)
transverse normal faults　横向正断层
trap confine　圈闭边界
trap formation　圈闭形成
true meridian　真子午线
turbidity current　混浊流,异重流
two sides of the same coin　一个硬币的两面
types of geological maps　地质图的类型

U(u)

ultimate oil recovery　油层最终采收率
unavoidable uncertainty range　不可避免的不确定度范围
under pressures　欠压
undersaturated oil reservoir　未饱和油藏
uniformity coefficient　均一性系数
unsampled regions　未采样地区

V(v)

various correlations　多种相关性
velocity pull-up　速度上拉
vertical joint　垂直节理
vertical movement　垂直运动
vertical resolution　垂向分辨率
vertical upthrust　直立逆断层
visual thinking　视觉思维
vitrinite reflectance　镜质组反射率
volatile plumes　挥发性烟缕
volcanic eruption　火山喷发
volcanogenic rock　火山岩

W(w)

water drive　水压驱动
water flooding　注水
water production　产水
water wet　亲水的
wave base　波基面,浪基面
wave migration　波迁移
wave trace　波形
weathering profile　风化剖面
West Siberian　西西伯利亚
wildcat well　初探井
wireline logs　电缆测井
with reference to　关于,根据